Blue Ridge China Traditions

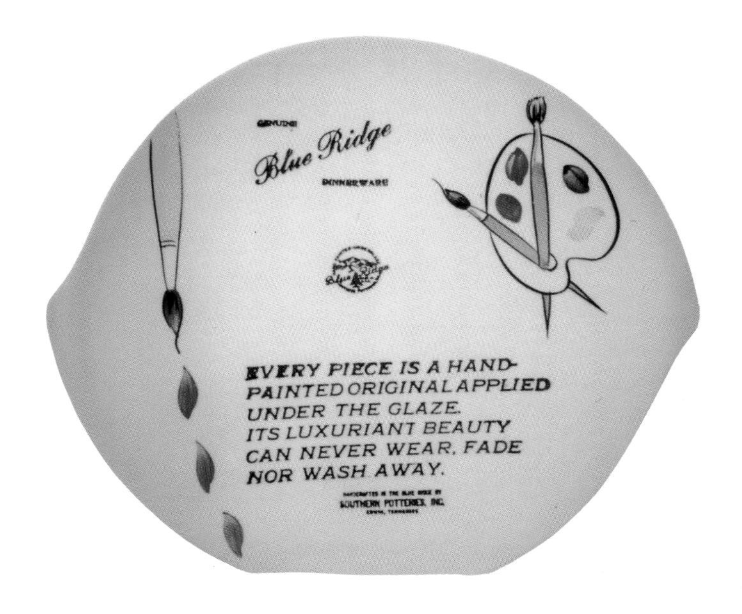

Frances and John Ruffin

4880 Lower Valley Road, Atglen, PA 19310 USA

Dedicated to the nicest people in the world, those whose lives have been touched in some way by Blue Ridge China.

Acknowledgments

We are happy and proud to present to you *Blue Ridge China Traditions*. Like *Blue Ridge China Today*, it is a celebration of the beauty and uniqueness of Blue Ridge China, a tribute to the talented people who created it, and a comprehensive guide for those who appreciate and collect these time worthy creations.

It is true that one of the best things about collecting Blue Ridge is the people you meet. Our lives and our book have been enriched through the contributions of many people. We enjoyed visiting and photographing several collections and appreciated the photographs sent to us by other collectors. Several descendants of pottery employees and former employees, themselves, were es pecially helpful in obtaining historical information. Blu Ridge collectors were quick to respond to our request for photographs of their pieces. Once again, our famil encouraged and aided us in this endeavor. Our friend were patient and understanding when the book took pre cedence over social events. To each of you, who we ac knowledge here, and to our anonymous contributors, w extend a sincere and heartfelt "Thank you!"

Sincerely
Frances & John Ruffi

Library of Congress Catalog Card Number: 99-60441

Book Design by Anne Davidsen
Type set in Seagull Hv BT heading font/Aldine 721 Lt BTtext font

ISBN: 0-7643-0822-X
Printed in China
1 2 3 4

Published by Schiffer Publishing Ltd.
4880 Lower Valley Road
Atglen, PA 19310
Phone: (610) 593-1777; Fax: (610) 593-2002
E-mail: Schifferbk@aol.com
Please visit our web site catalog at
www.schifferbooks.com

In Europe, Schiffer books are distributed by
Bushwood Books
6 Marksbury Rd.
Kew Gardens
Surrey TW9 4JF England
Phone: 44 (0)181 392-8585; Fax: 44 (0)181 392-9876
E-mail: Bushwd@aol.com

Contents

Contributors 4
Introduction 5
Chapter 1: History of Blue Ridge China 7
Chapter 2: Clinchfield China 17
    Advertising Pieces 25
    Granny Bowls 29
Chapter 3: Floral and Foliage Patterns 41
    Clinchfield and Astor Shapes 41
    Colonial Shape 51
    Accessory Pieces 77
    Candlewick Shape 89
    Piecrust Shape 101
    Skyline, Trailway, and Palisades Shapes 106
    Woodcrest Shape 122
Chapter 4: Fruit and Vegetable Patterns 124
Chapter 5: A Potpourri of Patterns 146
Chapter 6: Special Patterns 155
    Designer Series 155
    Feathered Friends 156
    Caribbean Series 160
    Artist Signed Pieces 161
    Character Jugs 164
    Country Life Series 166
    Countryside Series 167
    Provincial Line 174
Chapter 7: Seasonal Patterns 178
Chapter 8: Cookware 183
Chapter 9: China for Children of All Ages 188
    Demitasse 188
    Dinnerware 203
Chapter 10: Novelties 208
    Figurines 208
    Lunchtime Pieces 210
Chapter 11: The Tradition Continues 212
    Erwin Today 212
    Spin-Off Potteries 215
Appendix 227
    Backstamps 227
    Price Guide 231
    Pattern Numbers 234
    Glassware 237
    Resources 238
    References 238
Index 239

Contributors

Bill & Renee Accord
Jeff Adler
Charles & Jean Adams
Joyce & Paul Arbaugh
Steve Avery
Deward G. Bailey
Virginia & Lincoln Barber
Sandra Barefoot
Bill & June Basco
Jan Bates
Paul & Ruth Ann Batta
Judith Baughman
Robert & Sandra Beck
Blue Ridge Appreciation Group
Blue Ridge Café
Blue Ridge Pottery Store
Blue Ridge Pottery Club
Paul & Judy Boehm
Janise Bonds
Eunice & Carl Booker
Larry Boxum
Sabre L. Brayton
Lois Brazina
Lee Broyles
Shane & Kristy Buchanan
Karen L. Buckley
Bullrun Unlimited (Diana Bullock & Robert C. Runge, Jr.)
Doris Shipley Callaway
Barbara & Billy Jack Campbell
Vicki Carr
Glenn & Willene Clark
James Coleman
Michael & Marie Compton
Mr. & Mrs. Warren Congdon
Mildred & Jack Conley
Arlene Cooley
Betty J. Cox
Ron Curl
Kathy Deich
Becky Denny
Les & Ann Duncan
Kathy Dougan
Gary M. Edmonds
Charles Edwards
Candy Ellison
Maxie English
Erwin Chamber of Commerce
Erwin Public Library
Erwin Record
Martha Erwin
Mary & Ray Farley
Adrienne Felderman
Sheila Ferguson
Duane & Gail Fielder
Terri Ann Frantzen
Richard Freed
Mrs. Katherine B. Futrell
Karen Gerlach
Ruth Cox Goodman
Emily & Jerry Gordon
Kay Hackett
Peter & Mary Hamchuk
The Hanging Elephant Antique Mall
Shirley Hanson
Wanda & John Hashe
Marion Henry
Heritage Museum
Jeannie Heskett
Lori Hinterleiter
Madelyn Kimmel-Holley
Ward & Wilma Howard
Debi Howey
Betty W. Irwin
Amy Irwin
Janice Jones
Teri & Tom Jones
Just 1 Hour Photo & Custom Frame Shoppe
Winnie Keillor
Wanda Kellar
Ralph & Anneliese Kemp
Rene' & Danny Keplinger
Tara Kindschi
Margaret Kirkland
Terri Lawley
Judy Leach
Glenna B. Lewis
Joey Lewis
Judy E. Lewis
Lisa Lingerfelt
Sandy Lingerfelt
Helen Linville
Main Street Mall
Jacqueline Malone
Edythe Manfield
Pam Mann
Cheryl Martin
Wendy MacDonald
Iris McCain
David L. McCulley
Dave & Eloise McGinnis
Warren McIntosh
Ron & Cindy Melcher
Daniel K. & Mary Miller
Dorothy Miller
Jeffery Moss & Elizabeth Sterling
Judy Murray
Teresia Murray
Brenda Myers
Robbie Nelson
Pat Newbill
Sydney Nichols
Harvey & Betty Norris
Christy Parker
Jay Parker
Katie Pcenicni
Perry Antique Mall
Negatha & Earl Peterson
Wayne & Pam Phelps
Gary & Michele Ray
Morgan (Bucky) Redwine
Dana, Roby, Elizabeth, and Roby West Redwine
Lyle & Linda Rhodebeck
Jim & Irene Robertson
Vonda Rodriguez
Edna M. Rogger
Connie & Fran Rose
Parks & Karen Rountrey
Cheryl Ruffin
Lady Ruffin
Lisa & John Ruffin, Jr.
Ruby Runion
Robert M. Sabo
Karen Seers
E.C. Sellors
Doug Shank
Julie Shellhammer
Cliff & Mary Shelton
Dr. Betty Shepherd
Anita S. Sibrans
Dorothy Smith
Larna P. Smith
Linda Smith
Kim & Bryan Snyder
Southern Potteries Employees & Families
Debbie Spencer
M. Spolarich
Stangl/Fulper Collectors Club
Patty Bowen Stapleton
Dennis & Debra Stevenson
Larry & Carol Sutherland
Gerald & Nellie Ruth Taggart
Gail E. Taylor
Lou Thornberry
Tri-State Antique Center
Gwen Turner
U.S. Postal Service
Hazel McIntosh Vaughn
Vance & Ami Vogeli
Betty Wagers
Jamie Walls
Andrea, Scott, and Maggie Lee Weir
Larry Whittaker
Michael Dick Williams
Furman Willis
Jami Willoughby
Karen Parker Willoughby & Parker Girls
Wanda Vannoy Woods & Bob Woods
Betty Yates
Dolores R. Zopf

Introduction

We were overwhelmed and overjoyed with the success of *Blue Ridge China Today.* When we were asked to compile another book we were flattered and elated, but concerned over whether we could do an encore. Seeing *Blue Ridge China Today* widely used at sales and auctions, as well as cited on the Internet, was very exciting and rewarding. Therefore, we approached *Blue Ridge China Traditions* with the intent that it, too, would be appealing, informative, and efficiently useful. When combined with *Blue Ridge China Today*, they would reflect a true representation of Blue Ridge collecting.

The format of this book is basically unchanged from its predecessor, because it had been proven to be systematically correct. Chapter 1, History, has been enriched through the addition of remembrances and pictures from pottery employees and their descendants. Chapter 2, Clinchfield China, has been expanded due to a growing interest in advertising pieces and granny bowls. Many new or previously unpublished patterns have found their way into the gardens of Chapter 3.

Because of a noticeable quest for combination fruit and flower patterns, we began Chapter 4 with these colorful patterns. With the addition of some unusual patterns, Chapter 5 is truly a potpourri. You can see examples from the Caribbean, Mandarin, and Mexican series among the Special Patterns of Chapter 6. Chapter 7 offers some ideas for using your Blue Ridge in celebration of the seasons. We were delighted with the new finds of known patterns in cookware and think you will be pleasantly surprised and pleased with Chapter 8. Blue Ridge collectors are very resourceful. We noticed the use of Ovide coffeepots with Demi cups and saucers quite often. We were delighted that cake plates and even 10 inch plates were being used in lieu of demi trays. Interestingly, none of our research revealed that any demi trays were made and marketed as such. In fact, demi teapots and demi cups and saucers were sold as individual or after dinner sets with no mention of a tray. So congratulations Blue Ridge collectors on your ingenuity and creativity. Also in this chapter, a fairy tale came to life when Ruth Cox Goodman recognized the frog stamp shown in *Blue Ridge China Today* and remarked to her mother that she had that plate. Check out Chapter 9 for the rest of the story!

Chapter 10 begins with an adorable assortment of figurines, which haven' t quite been accepted into the Blue Ridge family. It ends with an unusual selection of "lunchtime pieces" that we know you will enjoy. Chapter 11 updates you on our beloved Erwin and the wonderful people who live there. Here you will also find examples from several "spin-off " potteries that continued the tradition of Southern Folk Art. These include examples from Erwin Pottery, The Cash Family Pottery, Clouse Pottery, Marie's, and unidentified pieces. Because accurate pricing requires extensive investigation into many areas, we have not attempted to suggest prices for the products of the spin-off potteries or the unidentified pieces.

The expanded appendix should be even more useful to the collector. It was not possible to repeat all the backstamps shown throughout *Blue Ridge China Today* and *Blue Ridge China Traditions*, but we have included several in the appendix. Thanks to Ray and Mary Farley (and many of you), our pattern number list is significantly more comprehensive. Thanks to Michael and Marie Compton, we have added a glassware list.

Since Blue Ridge collectors enjoy meeting other Blue Ridge collectors, we are working on a state-by-state list of groups or clubs. We were happy to hear the Blue Ridge West Luncheon will be held annually, with the June 1999 event already scheduled. There is also talk about a Blue Ridge South Show. We would like to hear from you with news of Blue Ridge events in your area as well as any Blue Ridge topic of interest to you.

It seems evident through your communications that you wish us to continue updating you on information and new patterns. Our addendum of August 1997 was a somewhat limited sample of what we would like to offer on a quarterly basis. Our plans for this type of publication will be finalized shortly and if you are on our mailing list you will be notified when it is a reality.

We were surprised at the number of new, or previously unpublished, patterns that have been discovered. As has been our policy from the beginning, we assigned pattern names only after extensive research into all published sources did not reveal one. The names we have assigned or discovered are enclosed in quotation marks. Following the precedent we set in *Blue Ridge China Today* and continued in our August 1997 Addendum, when patterns were similar to an existing pattern, we have indicated this by adding "variant" or designating an additional color. All names have been cleared through our data base of over 2,000 pattern names that have already been used, some more than once. We strive to notify you whenever a name has been used for more than one pattern. In these instances,

15" Platter with turkey decal, unmarked. Inscribed: Atlanta Furniture Co., Home is what we make it, 152 Whitehall St. N. W., Atlanta, Georgia.

we may add a "clarifying" word to the existing name as we did with "KIND OCTOBER". (October had already been an assigned name.) We endeavor not to duplicate names or rename patterns, but we do make every effort to use the pattern names assigned by Southern Potteries, jobbers, distributors and merchants.

As with *Blue Ridge China Today*, we authenticate the pieces we show and are not hesitant about rejecting a questionable piece or showing it as an unknown. The unmarked turkey decal platter pictured above is an example of one phase in our process of determining new patterns. Although it had the look and feel of an early Clinchfield advertising piece, it just wasn't quite right and our research did not support it as genuine. Therefore, we have not shown it as a product of Southern Potteries.

Our focus for *Blue Ridge China Traditions* was on new patterns or pieces. We have also repeated all the patterns from our Addendum of August 1997, since this was received only by those persons on our mailing list.

As in *Blue Ridge China Today*, a suggested price appears below pictures of most pieces. This price represents a midpoint within a range of today's prices. Keep in mind that this price is for a piece in mint condition, even when the photographed piece is otherwise. The price guide has been completely reevaluated and revised. We have considered Internet prices along with show, store and mail order prices. As many of you know, sales over the Internet continue to experience tremendous growth and have literally brought the antique store into your home via the technology of the computer. We also realize that many collectors prefer to shop the old fashioned way. Most importantly, remember that the price you pay, or ask, should be one that you can live with.

We are not infallible, but neither is our research limited or careless. We invite information as well as suggestions that will help us in our endeavor to present to you, the collector, the most comprehensive and current guide for collecting our beloved Blue Ridge!

Chapter One
History of Southern Potteries and Blue Ridge China

he Origin (1916)

When the Carolina, Clinchfield and Ohio (CC&O) ailroad began casting about for avenues to enhance its il shipping business, the idea of organizing a pottery ecame a reality with the establishment of the Clinchfield ottery in Erwin, Tennessee, in 1916. At the time, Erwin id less than 300 residents but was centrally located from e standpoint of the railroad for transporting the raw aterials for a producing pottery. The railroad had ear- er established machine shops in Erwin, which are still ilized by the CSX Railroad, the CC&O's successor.

The original pottery workers were lured from producing otteries in Ohio by E.J. Owens, who owned a pottery in ebring, Ohio. The original "Northern People," as they were rmed by residents of Erwin, trained those hired from Erwin id its environs for employment in the pottery.

They were housed in "pottery houses," which had been esigned by Grosvenor Atterbury, a New York architect, id constructed by the Clinchfield Railroad's affiliate, e Holston Corporation. These houses—there were ree basic floor plans—were rented to the Ohio potters r many years and still stand today.

Commercial production started in 1917 and was pri- arily centered around the production of standard din- erware using common chinaware molds and decal-ap- ied patterns. Southern Potteries became the official ime in 1920, when a corporate charter was obtained id $500,000 in stock was sold to the public.

The Early Years (1920-1938)

The Clinchfield crown backstamp was commonly seen on tableware sold throughout the United States from 1920 through 1938. Limited use of lining, an application of thin lines and borders painted on the china, was seen on some of the early patterns. For the most part, designs used during the early years were lifted from sheets of decals and applied by hand.

The Handpainted Era (1938-1957)

Charles Foreman, an associate of E.J. Owens, is generally credited with the introduction of hand painted designs to the production line around 1938. The earliest hand painting consisted of accent strokes and simple designs around the decals. Gradually, the decals were totally eliminated and all designs were painted by hand. Generally, a design artist painted the design pattern, which was then copied by the artists on the production lines. Frequently, a pattern was created by copying the designs of other potteries. Southern's Nove Rose pattern, along with several similar patterns, was derived by copying a design from a foreign pottery's platter. The French Peasant, Normandy, Brittany, and Palisade patterns were copied from the designs of Quimper, the famous French pottery. Copying patterns of others was a common practice throughout the life of Southern Potteries as it was for most of the potters of this era.

uthern Potteries Inc., Erwin, Tennessee, April 1925. Notice the bottle kilns over the roof line.

Southern Potteries was the largest producer of handpainted china during its primacy in the early 1940s to the early 1950s. After the Second World War ended in 1945, a gradual erosion in overall production occurred due to escalating labor costs, competition from Occupied Japan, and the advent of plastic dinnerware. Finally, the pottery was forced to close its doors in early 1957 after months of layoffs and reduced workweeks proved unsuccessful in keeping the pottery alive. All but a handful of the remaining workers were notified that they were to be terminated as of February 28, 1957, since the stockholders had voted on January 29 to liquidate the corporation. Stockholders were paid a liquidating dividend, indicating that the management was fiscally very conservative but pragmatic in their view of the future. In spite of their optimism, Southern Potteries closed its doors permanently in 1957.

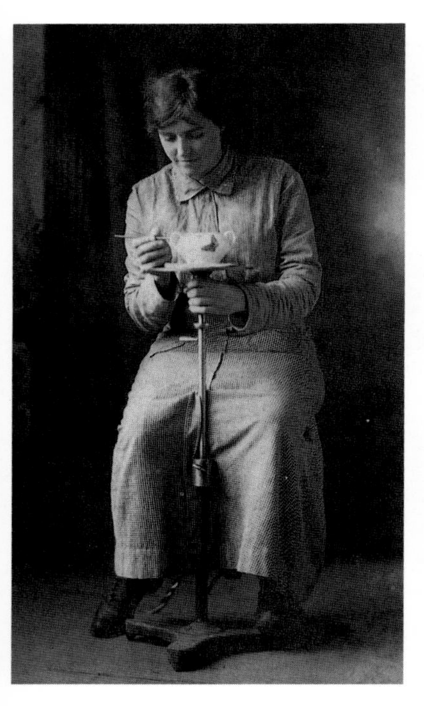

Novella Beals practices "lining" in February 1917 in advance of the pottery's opening. Early painting was limited to lining of decaled pieces.

James White positions the bisque for firing (1933).

Mallard Hilton, like so many of the potters, came to work as a kiln placer directly from the farm (1933).

Walter Hensley, head kiln fireman, stokes one of Southern's coal fired kilns (1933). Hensley gave up a job as foreman because he refused to work on Sundays. He worked at Southern from 1917 until he died in 1952.

Paul Bowen, who worked at the pottery from 1939 to 1957, came to Southern from the Paden City Pottery. He became production manager and was one of the few historians of the pottery. He was an amateur photographer and recorded much of the surviving photographic documentation of the pottery.

Southern Potteries employees and their families at the annual company picnic in Asheville, NC (1946).

Filling orders in the warehouse.

oris Shipley Callaway, shown in a 1947 photo-
aph with her husband James. Doris partici-
ıted in a beauty contest held at the company
cnic in 1946.

Above and Right: Fermon and Warren McIntosh were both dishmakers. They made the tools used in crafting the pottery from raw clay.

Southern Potteries maintained eleven sales rooms around the country and from these accepted orders from buyers for both large and small retailers and jobbers.

The Workers and Artists of Southern Potteries

While there were some very talented artists and designers employed by the pottery throughout its history, the bulk of the production work was done by teams of painters working in groups of three or four. The lead artist would paint the main object of the design, and then others would add detail such as leaves, stems, and buds to complete the design. Limited numbers of patterns were painted from start to finish by a single painter. Among these were the artist signed plates and platters as well as the Betsy and character jugs.

Designs for some patterns were obtained from art students at colleges. An example of one such pattern was Salmagundi, which was the design of Kay Hackett, a student at Alfred University in Aurora, New York. Kay's design was produced by Southern Potteries in 1941. Kay went to work at Stangl Pottery as a designer after graduating from college and became one of their most respected artists.

Southern Potteries created its products from raw ingredients, which were mixed by workers at the pottery. The raw materials such as feldspar, flint, kaolin, and coal were transported by rail to Erwin and much of the production also left by rail. The bisque from which the various shapes were fashioned and the glaze used to provide a permanent seal for the handpainted designs were mixed by Southern's potters. There was very little in the way of modern machinery utilized by Southern during its history. Most of the production was done by hand, the old fashioned way, which caused production to be very labor intensive.

While the products of the Clinchfield and Southern Potteries were never considered to be destined for the finest tables, there was always an effort to stay abreast of the current trends in everyday dinnerware and accessories. Most of the sales occurred through eleven Blue Ridge showrooms throughout the country, at major department stores, and through catalog sales and premium offers. Many of the pottery's seconds were sold to distributors, who in turn sold to smaller retailers, primarily in the southeastern states.

The Spin Offs

A number of Blue Ridge artists and pottery employees formed their own potteries after leaving Southern Potteries. While each has a rather distinctive appearance, there are still similarities to the Blue Ridge style of Southern Potteries.

Production workers grade and sort finished goods in the warehouse.

Erwin Pottery

Negatha Peterson and her husband, Earl Peterson, are still active in their business located on Main Street in Erwin. Negatha, a former Blue Ridge artist, has created a following in the world of cookie jars and her works are highly prized and much sought after by collectors. Additionally, she has a significant collector base for the many items she produces, using the original Blue Ridge molds and patterns.

Cash

Probably the most prolific of the early spin offs, the Cash family produced Blue Ridge-like pottery beginning in 1945. Their works have gradually increased in value and collectors have begun to find it difficult (and expensive) to find some of the more desirable pieces.

Marie's

Marie Branham is still producing Blue Ridge style patterns on shapes that are not traditional for Blue Ridge. Her work seems very precise and is appealing to the eye of many Blue Ridge collectors—and to others who like handpainted pottery. Her pottery is located in Rockwood, Tennessee.

Clouse (Unaka)

Some of the finest detail among the former Blue Ridge artists is found in the works of Bonnie Clouse. Bonnie's sister, Lena Watts, helped her to make Clouse items some of the most beautiful examples of Blue Ridge techniques and patterns

arehouse workers at a Christmas dinner. Front row (Left to ght): Pearl Bailey, Florence Strickland, Belle Hensley, Dale wis, and Joanne Harris. Back row (Left to right): May Stancil, rgie Bailey, Viola (Dode) Foster, Lillian Tolley, and Nora Smith.

Christmas party for pottery workers on December 9, 1948.

ecollections of Southern Potteries

Warren McIntosh

Warren McIntosh, a dishmaker for Southern, recalls vidly the process followed for molding the raw clay into e shapes for painting and firing. The following descrip- on is provided in his own words:

he clay is brought to the dishmaker in rolls about 30"- " long and about 6" in diameter. The dishmaker kes a piece of clay off the roll and spreads it out to a ickness of about 1/4" using a batter. This process is lled batting out. After flattening to the thickness and ape needed, the dishmaker uses a polishing knife to n over the clay to take out all rough spots. He then aces the flat piece of clay on the mold. Next, he lifts the edge of the clay on the mold to get rid of all air om under the clay. Using the dishmaker's pitcher, he es over the bottom of the dish or baker to make it nooth. Next he uses a sponge and "runs" the air from der the clay. Next, he takes the dishmaker frog and ts the excess clay from the mold and forms the edge the dish or baker. Next the sides of the dish or baker e formed and smoothed using a small rubber tool. he dish or baker is then placed on a rack to dry for out one hour. Then, it is removed and the sides and ttom are polished. It goes back on the rack until it is mpletely dry. It is then removed from the mold by e finisher and all rough edges are removed and all arks are sponged from the ware. A frog is a slingshot pe of tool, which uses wire as a cutting edge. The wire as usually taken from old automobile tires by burning em to make it easy to remove the wire. The wire was en tied over the frog to form the finished dishmaker's og. The pitcher is made from ceramics, and the bot- m of the batter is made of plaster of Paris. The tools the dishmaker were for the most part handmade by e workers.

Farewell luncheon in February 1957. Standing (Left to right): Alvin Miller, Webb Gentry, and Albert Price. Seated (Left to right): Louise Moore, Eva Miller, Ruby Randolph, Doris Miller, Juanita Perry, and Louise Miller.

Paul Bowen, Webb Gentry, Albert Price, and Alvin Miller share lunch time conversation about the pottery's closure.

Doris Shipley Callaway

"It was a beautiful summer day in 1946. In fact it was the day of the annual picnic of Southern Potteries, Inc. of Erwin, Tennessee. The setting was Buncombe County Recreation Park in Asheville, North Carolina. There was a lake and a roller skating rink. 'Coconut Grove' was playing on the jukebox as the skaters went around and around.

The main event of the day that really stands out in my mind was the beauty contest. This was publicized for some time, but little interest was shown. At the urging of friends, I finally agreed to be in it. There were only four of us: Ethel Smith, Sara Laughrun, Martha Baxter, and myself. We had to wear swimsuits. (I borrowed mine from one of my sisters. It was a two piece in a yellow floral design.) An elderly woman was one of those wanting me to participate in the contest. Later, she commented, 'I wouldn't have insisted if I had known you had to wear a bathing suit!' Anyway, we had to walk around the swimming pool four times. I was shy and so embarrassed. I didn't even look at the judges when we went by them, but down at the ground. I did come in third. Ethel Smith won and was named Miss Blue Ridge. Her picture was in the Johnson City paper. I was just glad it was over!"

Deward Bailey worked at the pottery until it closed in 1957. He fondly recalls the friendly atmosphere and playfulness of the workers. He also remembers the intense heat from the frit kiln, where he worked to make glaze.

Deward Bailey

Deward was a young man when he first started working at Southern Potteries. "I did a lot of jobs—everything from mowing grass to making glaze at the frit kiln," relates Deward. The job he enjoyed was carrying the ware boards from the painters to the glaze area. He knew and remembers all of the painters. "There was a lot of fun in working at the pottery. It was very sad when they announced that the pottery was closing. I had no job to go to." Perhaps the hardest job he did was running the frit kiln, which produced the liquid glaze used to spray or dip the bisque after it was painted. "It got real hot around that kiln," he recalls. There was always a lot of "joking and kidding around with coworkers." He recalls that he used seconds from the pottery for serving meals at his home.

Lou Thornberry

Another recollection of the company picnics came from Lou Thornberry, whose father worked as a jigger after World War I. He recalls that the picnics were a "grand time" for a small boy growing up around the pottery. Most of the employees and family members were driven to the picnics in buses. On the way back from the last picnic he and his mother attended, the bus hit a bridge railing, and the bus was left hanging over the Spivey River in south Unicoi County. He and his mother were sitting in the front seat of the bus and had a good view of their perilous situation. The driver was successful in backing the bus out of its predicament, and they returned safely to Erwin.

Southern Potteries loading dock in 1950.

Postcard showing the same view of the pottery as the Erwin Pottery platter.

Above: Advertising billboard on the Ashville highway.

Right: Late addition to the pottery—a metal storage shed.

ilroad siding with the spar mill in the background.

Parking lot at Southern Potteries in 1950.

Erwin Pottery artist's rendition of the pottery as it looked in 1916 shortly after construction.

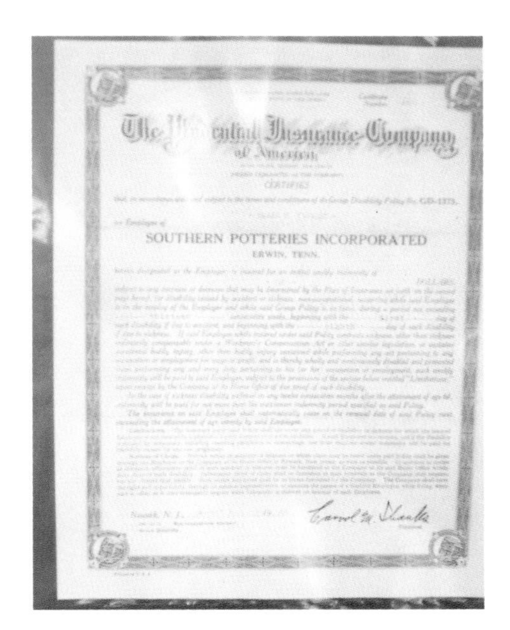

SOUTHERN POTTERIES INCORPORATED
ERWIN, TENN.

Southern Potteries, Inc.'s Group Disability Policy with the Prudential Insurance Company, 1951. The policy was for a weekly indemnity of $10.00.

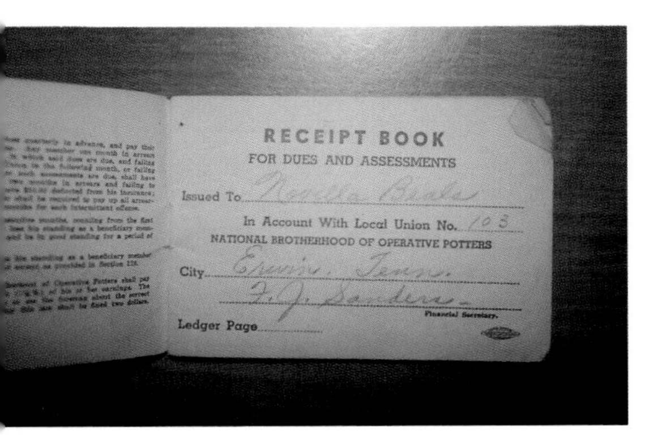

RECEIPT BOOK
FOR DUES AND ASSESSMENTS

Issued To Novella Beals

In Account With Local Union No. 103
NATIONAL BROTHERHOOD OF OPERATIVE POTTERS

City Erwin, Tenn.

F.J. Sanders
Financial Secretary.

Ledger Page

eceipt book for dues and assessments issued to Novella eals in account with the local union No. 103 of the ational Brotherhood of Operative Potters.

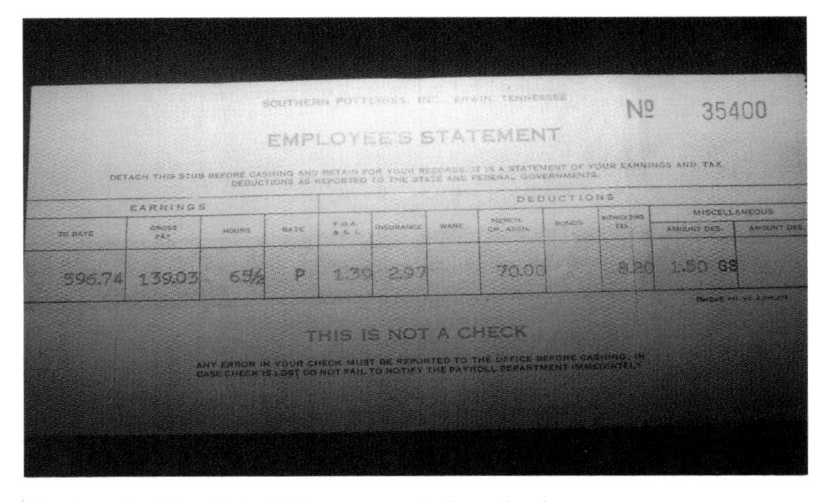

SOUTHERN POTTERIES, INC. ERWIN, TENNESSEE

№ 35400

EMPLOYEE'S STATEMENT

DETACH THIS STUB BEFORE CASHING AND RETAIN FOR YOUR RECORDS. IT IS A STATEMENT OF YOUR EARNINGS AND TAX DEDUCTIONS AS REPORTED TO THE STATE AND FEDERAL GOVERNMENTS.

EARNINGS				DEDUCTIONS						MISCELLANEOUS	
TO DATE	GROSS PAY	HOURS	RATE	F.O.A. & S.I.	INSURANCE	WARE	MERCH. CR. ASSN.	BONDS	WITHOLDING TAX	AMOUNT DES.	AMOUNT DES.
596.74	139.03	65½	P	1.39	2.97		70.00		8.20	1.50 GS	

THIS IS NOT A CHECK

ANY ERROR IN YOUR CHECK MUST BE REPORTED TO THE OFFICE BEFORE CASHING. IN CASE CHECK IS LOST DO NOT FAIL TO NOTIFY THE PAYROLL DEPARTMENT IMMEDIATELY.

Employee's statement which accompanied paychecks.

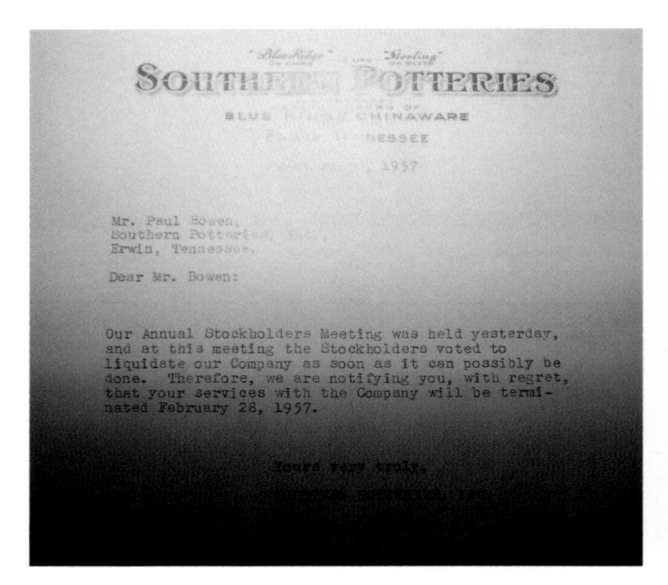

"Blue Ridge" ... "Sterling"

SOUTHERN POTTERIES

BLUE RIDGE CHINAWARE

ERWIN, TENNESSEE

1957

Mr. Paul Bowen,
Southern Potteries, Inc.,
Erwin, Tennessee.

Dear Mr. Bowen:

Our Annual Stockholders Meeting was held yesterday, and at this meeting the Stockholders voted to liquidate our Company as soon as it can possibly be done. Therefore, we are notifying you, with regret, that your services with the Company will be terminated February 28, 1957.

Yours very truly,

SOUTHERN POTTERIES, INC.

Termination letter to Paul Bowen signed by H.W. Kibler.

Southern Potteries, Local Union No. 103, Erwin, Tennessee. Float promoted a theme relevant today: "Be American, Buy American."

Belle and Walt Hensley were both employed by the pottery, as were large numbers of their relatives and friends.

Original Southern Potteries casings used to make the molds for the granny bowls.

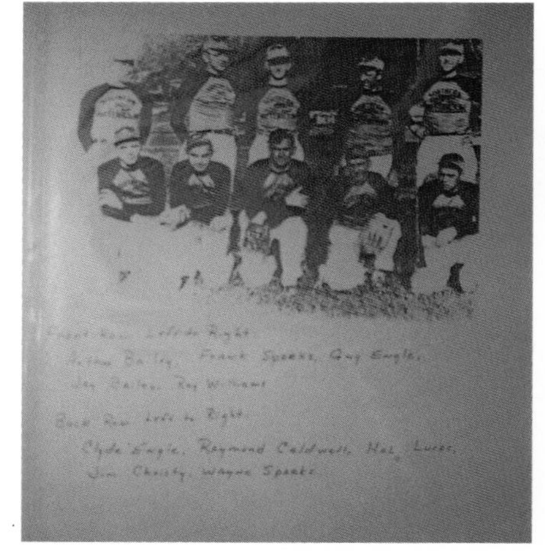

Southern Potteries' baseball team.

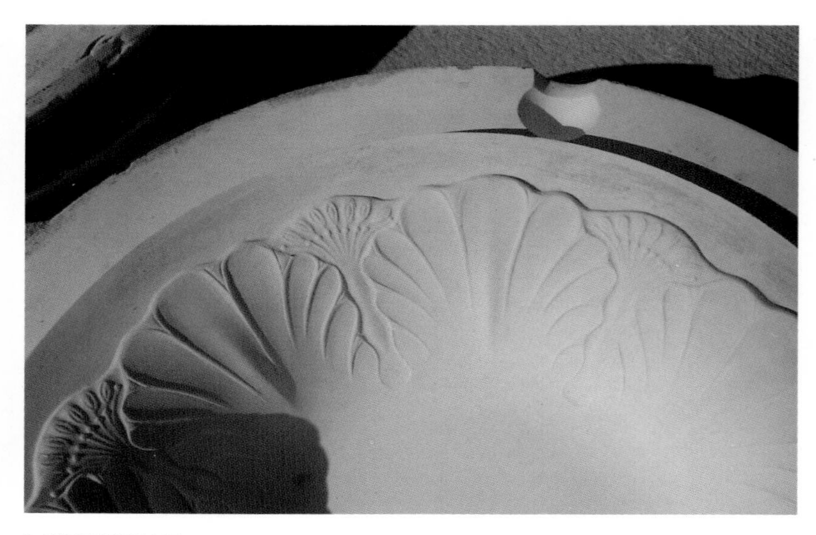

LOTUS LEAF

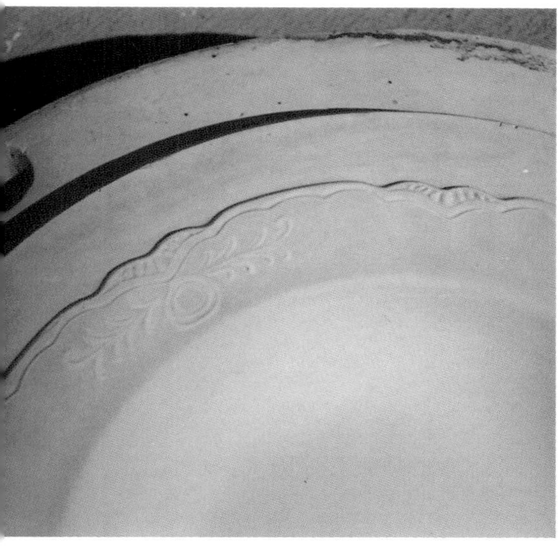

ROLL EDGE

LACE EDGE

RIPLE ARCH

FLORAL POINT

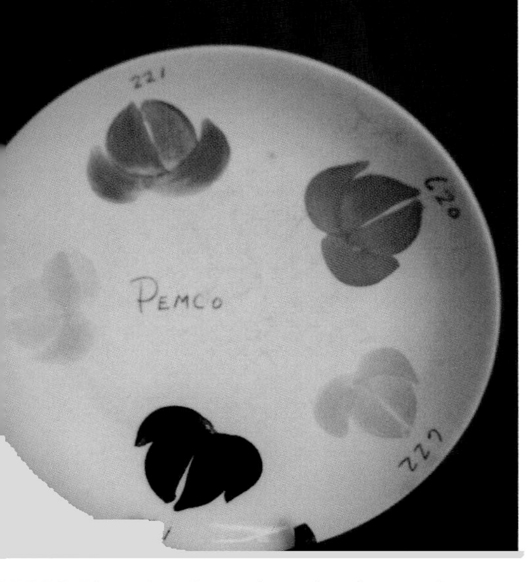

MCO Plate showing paint colors by number.

(Left to right) Short brush used for painting flowers, long brush used for leaves, long brush used for stems, shader brush for edges of plates, and a short brush for small flowers.

PEMCO Plate showing paint colors by number.

Glaze Sample Plate. Signed: Regular Glaze, Litho Dept. Each sample was numbered.

Pitcher tools used by Fermon McIntosh and Warren McIntosh to smooth bottoms of flat bottom pieces such as the BACHELOR BUTTON, Astor oval baker shown here.

Kay Hackett with SALMAGUNDI, the pattern she designed for Southern Potteries. *Courtesy of Bull Run Unlimited, Diana Bullock and Robert C. Runge, Jr. and the Stangl / Fulper Collector's Club.*

Brushes, burlap roller for WOODCREST, and jiggerman's tools.

Chapter Two
Clinchfield China

Clinchfield China has become a vital part of Blue
dge collecting. Advertising pieces have risen notably
price to meet the demands of collectors. Granny bowls
now a part of almost every Blue Ridge collection.

"AQUAMARINE DREAM," Sugar and Creamer Set, $50. Marked in gold: S.P.I., R.A., SAMML 31.

inchfield and Astor Shapes

chfield China shipping boxes are very collectible. "OLD ROSE," e pattern shown.

"Old Rose," Creamer, $25.

"TRIED & TRUE," Lustre, Console Set, $150. Outer rim of ruffle top vase matches lustre of base.

"GOLD LUSTRE," Teapot, $50.

"ROSARIO," Platter, $25.

CREST, Clinchfield, 12" Platter, $35.

FLOWER BASKET, Lustre, Teapot, Creamer, and Sugar, $100.

"GRANNY'S PLATTER," Handled Platter, $45. This platter has the old SPI backstamp.

"LUCINDA", Dinner Plate, $30.

"MARIGOLD," Watauga, Tea Set. Note handle on the creamer.

RST CLASS, Clinchfield, 9" Plate, $30. This d decorated plate has the Clinchfield crown kstamp and is dated 12-24.

"SCONCES," 3" Butter Pat, $40.

"ROYAL WREATH," Bread Tray. Inscribed in gold: Give us this day our daily bread.

"ROYAL WREATH," Dinner Plate, $25.

"MONOGRAM," Dinner Plate, $30. Salt and Pepper Shakers, $65. Monogrammed in gold, presumably these were special orders. Unlike later, more slender shakers, these early pieces are more rounded with fluted bottoms.

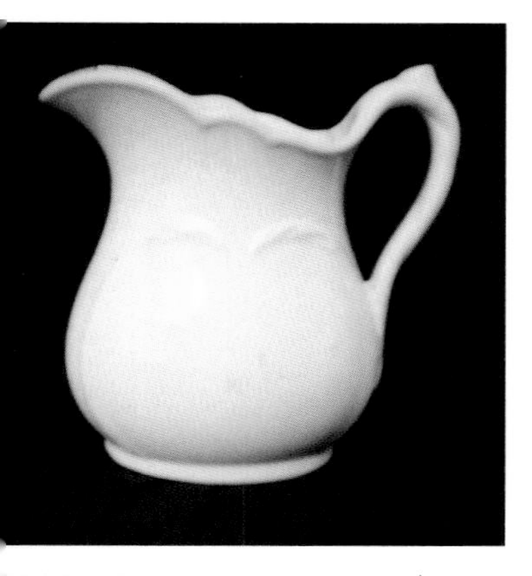

HALK," Lace Edge, Pitcher, $35. is is a very early piece.

"GOLD LEAF," Watauga (shape), Gravy Boat, $45.

"GOLD BAND," 7" Pitcher, $75.

"CHALK," Watauga, Pitcher, $60.

"PEACH POPPY," 11" Platter, $20.

"NOVELLA," Paneled Cup, with rose decal inside, $20.

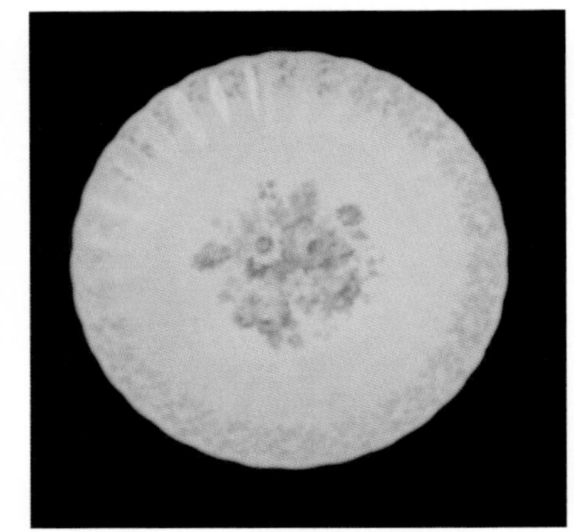

"NOVELLA," Wide Rib, Dinner Plate, $25. This may be an early Colonial shape.

"GOLD BRAID," Pitcher, $50.

"FLATLINE," Gravy Boat, $45.

"ROSE WREATH," 12" Platter, $20.

"NANA," Many gold-edged decal plates have the National Brotherhood Operative Potteries backstamp.

"ANTIQUE ROSES," Lace Edge, Bowl, $25.

"JAMI, " 12" Platter, $25.

[Q]UEEN," Watauga, 14" Platter, $100. The lady in this [de]cal platter is surrounded by what appear to be good [lu]ck symbols: the horseshoe, four-leaf clovers, and a [wi]shbone. The "backward swastika" is thought to be an [In]dian symbol for good luck.

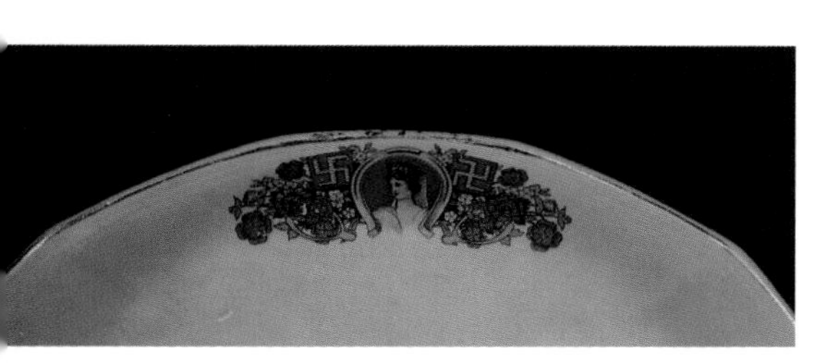

[C]lose-up of "QUEEN." Clinchfield crownbackstamp.

"ANTIQUE ROSEBUD," China Shakers, $75 pair.

POND IRIS, Watauga, Creamer, $40. This piece combines transfer decorating with hand painting.

Inside of POND IRIS Creamer.

"COLOR RING," 8.5" Square Plate, $25. Although this is an early handpainted plate, it is more colorful than later plates with similar patterns, such as GUMDROP WREATH.

Backstamp on "COLOR RING" is an old one: SOUTHERN HAND PAINTED CLINCHFIELD WARE, MADE ERWIN, TENN. U.S.A.

"LUSTRE GRAPES," 11.5" Platter, $25. Clinchfield Crown backstamp with 6-28 date.

BLUE WILLOW, Sugar and Creamer Set, $50. Cup and Saucer, $35.

"NEW WORLD," Alice Pitcher, $125.

"LUSTRE WILLOW," 8" Plate, $50. This may have been an early decorative piece, not part of a dinnerware set.

"HOLLAND," Clara Pitcher, $125. Unmarked but assumed to be made at Southern Potteries, although identical blue transfer pitchers have been found with a Made in Czechoslovakia backstamp.

OBERT E. LEE, Lace Edge, Plate, $100.

SANTA MARIA, Scroll Edge Bowl, $35. This decal was used by several potteries.

"PAINTER'S PAIN," Early Handled Lamp, $200.

OBERT E. LEE, 7" Square Plate, $150. Robert E. Lee cal pieces are collectible outside the field of Blue Ridge.

"DARK CHOCOLATE," Clara Pitcher, $100. Although unmarked, this piece has been assumed to be an early product of the pottery.

"SHEENA" Handled Vase, $125.

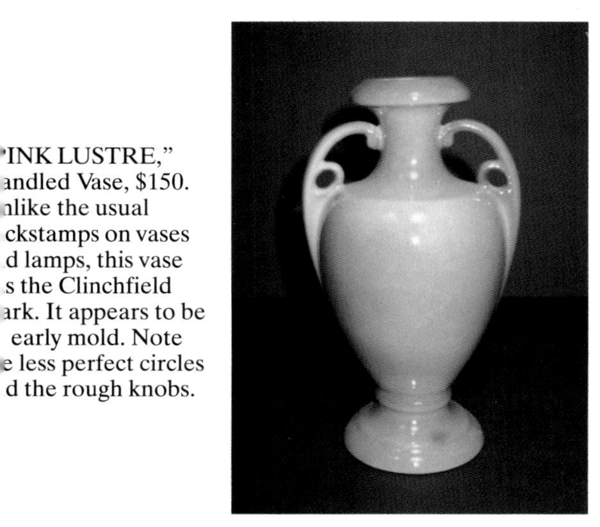

'INK LUSTRE," andled Vase, $150. nlike the usual ckstamps on vases d lamps, this vase s the Clinchfield ark. It appears to be early mold. Note e less perfect circles d the rough knobs.

"PEARL," Handled Vase, $100, Unmarked.

Above: ART NOUVEAU, Lady Flower Frog, $500. Clinchfield stamp.

Below: Bottom of Lady Flower Frog with Clinchfield backstamp.

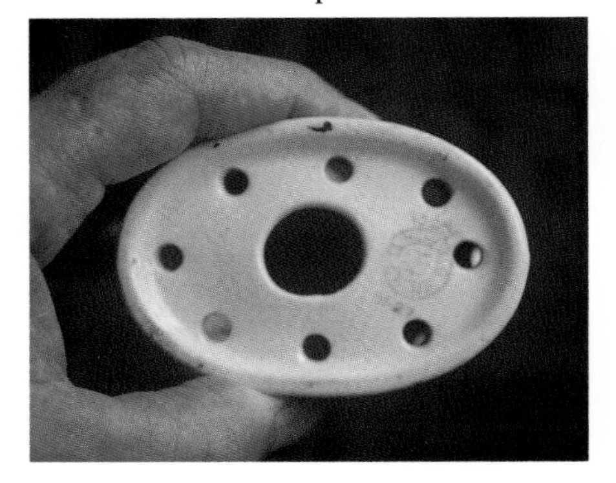

"HOOVER," Figurine. This elephant figurine was commissioned to commemorate Herbert Hoover's campaign visit to Elizabethton, Tennessee, in 1928. History was made when Southern Potteries, Incorporated was commissioned to manufacture pottery elephants to commemorate "Hoover Day in Dixie." Legend has it, through research and documentation by Dan W. Crowe, that Herbert Hoover, the Republican nominee for the 31st President of the United States, chose Elizabethton, Tennessee, for his only Dixie campaign site. Note that 1928 was a presidential election year, and his campaign was held on October 6, 1928. During this time, Elizabethton attracted national attention and became the biggest political extravaganza in the annals of East Tennessee.

"JUNE TULIPS," Clara Pitcher, $225.

The other side of "JUNE TULIPS" Pitcher, inscribed: "Compliments of R.G. Garrett, Montvale, VA."

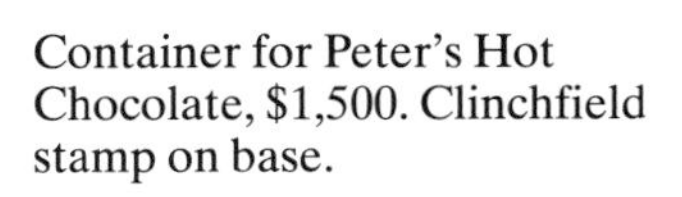

Container for Peter's Hot Chocolate, $1,500. Clinchfield stamp on base.

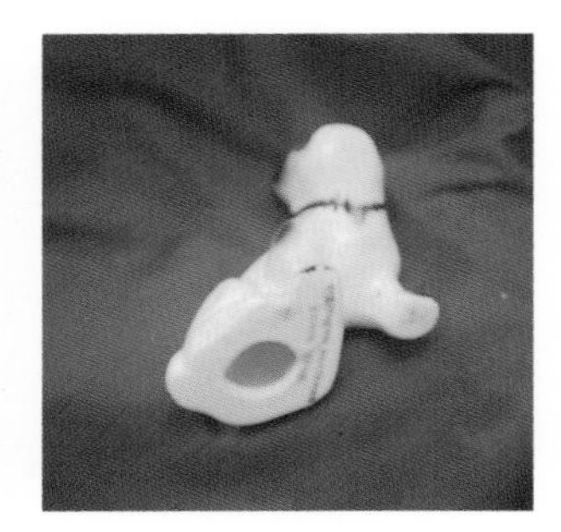

Left: "SPOT," Dog Figurine, $175. Inscribed on back: B.P.W.C.

Right: Backstamp on bottom of "SPOT." Stamped: Southern Potteries, Erwin, Tenn.

dvertising

Advertising played a large role in early Clinchfield roduction and continued to be a profitable aspect of ne pottery business. Ashtrays dominated production in ne beginning, but then gave way to dinnerware pieces, specially 9" granny bowls.

shtray, $45. Inscribed: "The eople's Bank, Johnson City, enn., Member F.D.I.C."

Ashtray, $40. Inscribed: "Clinchfield YMCA, 84th Anniversary, Erwin, Tennessee.

Clinchfield Railroad Co., Ashtray, $50. Inscribed: Fast Through Freight Service Between Central West and South Atlantic States.

Back of Clinchfield Railroad Ashtray with the Southern Potteries, Inc. backstamp indicating Warranted Gold 22 KT, Made in U.S.A.

HOMER," Bristol Twins Ashtray, $75.

"KATIE," Ashtray, $60.

Above: SUNGOLD, Ashtray, $60.

Left: SPRING GLORY, Ashtray, $60.

TULIP RING, Bristol Twins Ashtray, $65.

SHOW OFF, Ashtray, $60.

TWO OF A KIND, Ashtray, $100. Inscribed: "Compliments of Southern Potteries, Inc, Erwin, Tenn."

GILLYFLOWER, Ashtray, $75. Inscribed: "The United Potteries Co., Renkert Bldg., Canton, Ohio."

"PEGGY SUE," Ashtray, $65

SPIDERWEB (YELLOW), Ashtray, $35.

"SPLATTER ME BLUE," Clinchfield Railroad Co., Ashtray, $65.

REFLECTION, Ashtray, $55. This pattern is also called SUNBURST.

LAUREL BLOOMERY, Ashtray, $45. Many of the ashtrays with five indentations have Palisades patterns.

"VINCA VARIATION," Ashtray, $60. Clinchfield Railroad Co.

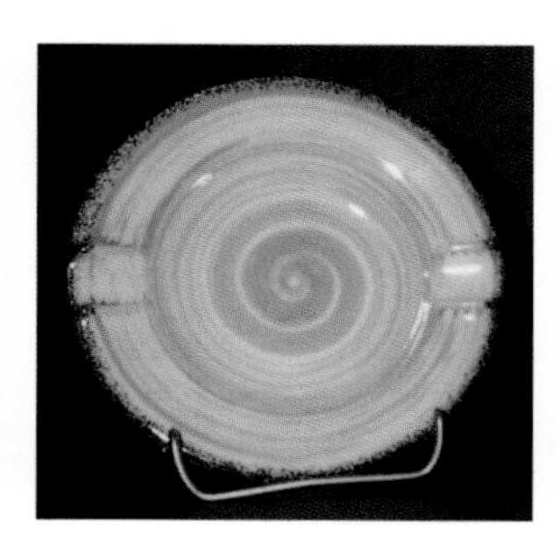

ROUNDELAY, Ashtray, $20.

STACCATO, Ashtray, spoon-rest style, $40.
STACCATO, Ashtray, $35.

ROWNVETCH, Ashtray, $40. This handled ıtray was also used for a spoon rest as shown and scribed on the Quaker Oats box. Today they are ɔre commonly used as spoon rests than ashtrays.

"SPLATTER ME BLUE," Antique Pitcher, $150. Inscription: "Compliments of Poages Mill Service Station, Poages Mill, VA."

"METALLICA BLUE," Ashtray, $55.

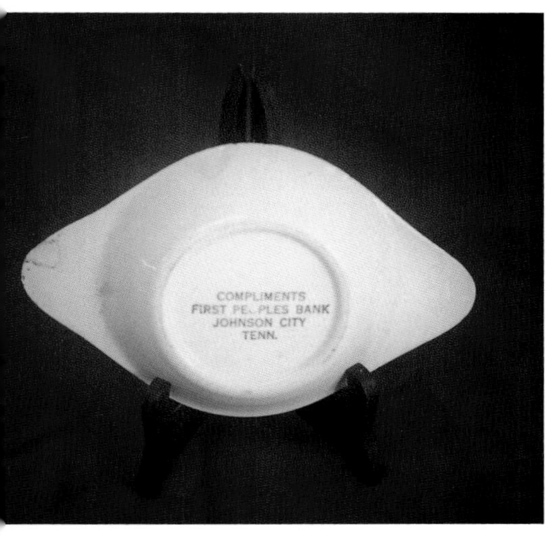

ck of the CROWNVETCH Ashtray. Inscribed: 'ompliments, First Peoples Bank, Johnson City, Tenn."

(Left to right) MOTTOE WITH SNOWFLAKE, Candlewick, Bowl, $45. Inscription: "Compliments of J.B. Stinson, Roanoke, VA." Square Rib Plate, $50. Inscription: "Pine Service Station, Floyd Star Route, Roanoke, VA." JANICE, Scroll Edge, Bowl, $25. Inscription: "Compliments of B. Borinsky, Princeton, W. VA." A large percentage of early Clinchfield production was advertising china. The Clinchfield Crown mark can be found on most pieces. Usually the advertiser wanted his name and often his wares or service on the front of the piece. Sometimes the inscription was put on the back, so the piece could easily be used in serving, without embarrassment.

OGO," Ashtray or spoon rest with Blue Ridge go. (Nice to see the logo on the front!)

"URN OF ROSES," Watauga, Vegetable Bowl, $35. The inscription on this bowl: "Compliments of R.N. Hudson, Mobile, AL."

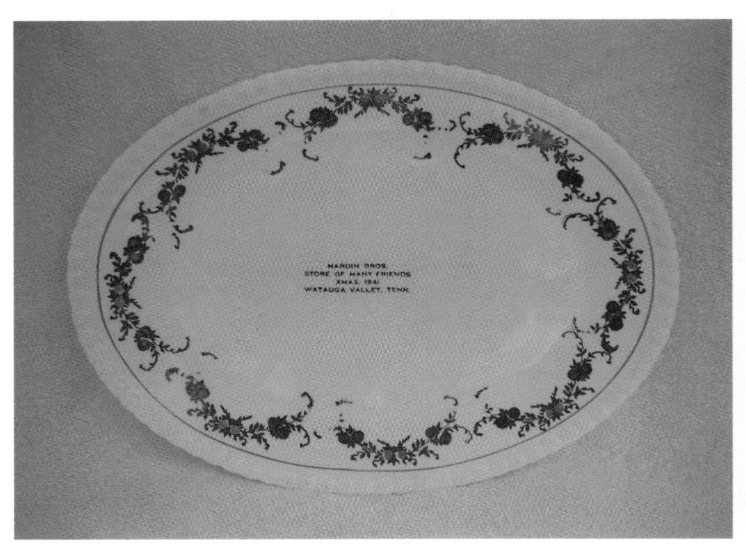

CATOSA, Candlewick, 11" Platter, $30. "Presented by Hardin Bros., Store of Many Friends, Xmas, 1941, Watauga Valley, Tenn." BLACK WIDOW is a similar pattern.

EDITH, Lotus Leaf, 9" Round Bowl, $25. Inscribed: "Kents Furniture Store, Tifton, GA."

Above: "HONEY BOUQUET," Honeycomb, 10" Bowl, $45.

Back of "HONEY BOUQUET" Bowl, Inscribed: "Compliments of H.G. Peters, Moneta, VA."

Above: "FIREFLY," Honeycomb, Bowl, $25.

Back of "FIREFLY" Bowl. Stamped: "Compliments of A.C. Hogert, Jeweler, Leaksville, N.C."

Copps Super-Market advertisement from the Johnson City Press Chronicle dated Thursday/ Friday, May 5-6, 1949. Note the heading: "Next to Southern Potteries."

TRIFLE, Candlewick, 9" Bowl, $75. Inscription: "Kents Furniture Store, Tifton, GA."

"ATHENA," Curlique Arch, 7" Bowl, $30. Usually the Scroll edge is found on 9" bowls.

YPSY DANCER, Candlewick, Adver-
ing Plate, $40. Inscribed: "Compli-
nts, A.L. Addington, Give us this day
r daily bread, Kingsport, Tenn."

MIDSUMMER ROSE, Wide Rib, 9" Bowl, $35.

"BLUE BUTTERFLY," Colonial, 9"Bowl, $45. "Presented by Kents Furniture Store, Tifton, GA." Also found on Lace Edge shape with a blue edge.

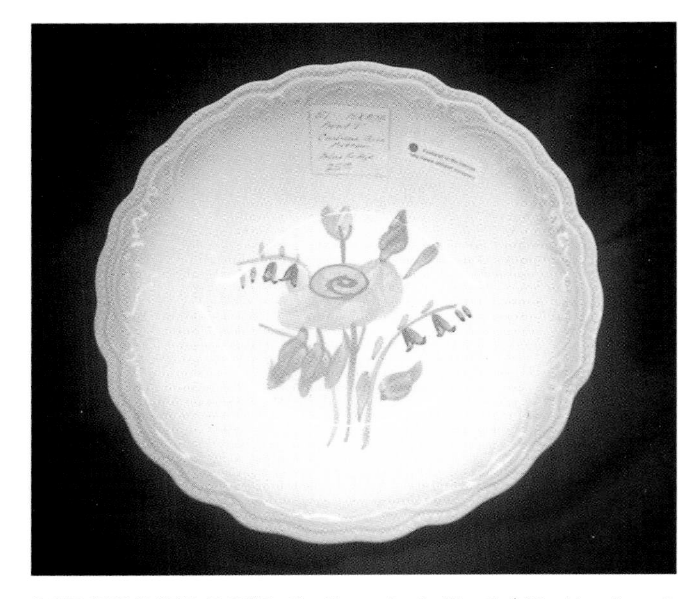

MIDSUMMER ROSE, Curlique Arch, Bowl, $30. Also found on green Monticello and light green Colonial shapes.

MIDSUMMER ROSE, Honeycomb, Bowl, $45.

NAÏVE, Lace Edge, Bowl, $25.

(Left to right) "MATILDA," Floral Point, Bowl, $25.
FINESSE, Wide Rib, Bowl, $25.

GYPSY FLOWER, Scroll Edge, 9" Bowl, $20.
Found with a variety of color borders.

Back of GYPSY FLOWER Bowl, indicating
Pattern #1542-UC.

"DODE," Curlique Arch, 9" Bowl, $25. This bowl is stamped with the "CROSSBAR" pattern.

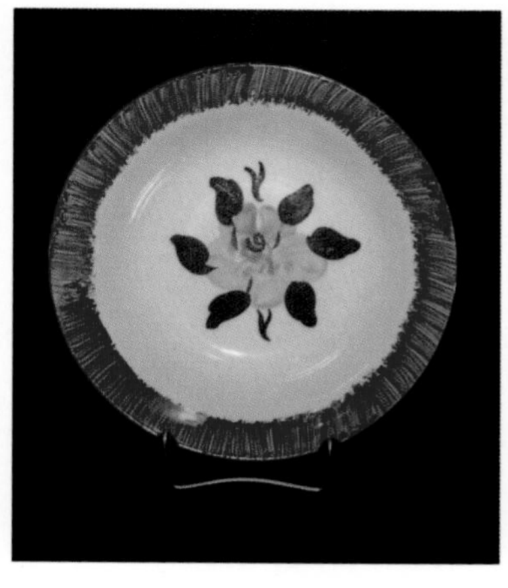

"SPLASHY SOUTHERN ROSE," Bowl, $30.

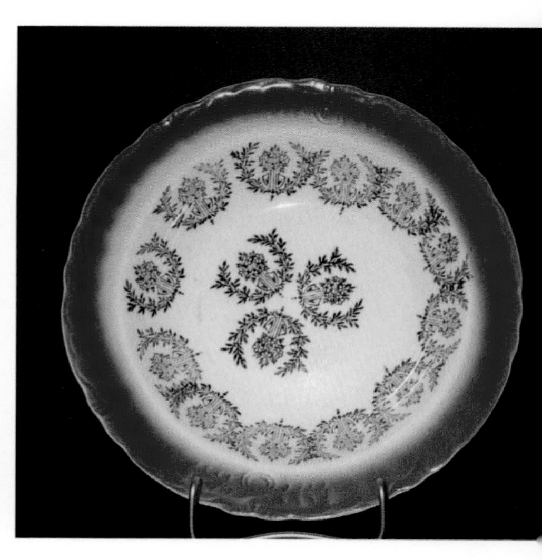

"MATILDA," Lace Edge, Bowl, $25.

"GRACE MARIE," 9.5" Bowl, $25.

"TARGET," Floral Point, Bowl, $25. "LACY RADIANCE," Lotus Leaf, 10" Bowl, $35.

AM," Scroll Edge, Bowl, $25.

"THE TREE," Lace Edge, Bowl, $45.

CLOVERLAWN, Scroll Edge, 9" Bowl, $25.

NIGHTLIGHT, Floral Point, Bowl, $25. "CROWNFLOWER," Honeycomb, Bowl, $30.

VICKY (Blue), Lotus Leaf, Bowl, $35. Also light green and pink.

"HERE'S LOOKIN' AT YA," Scroll Edge, Bowl, $55.

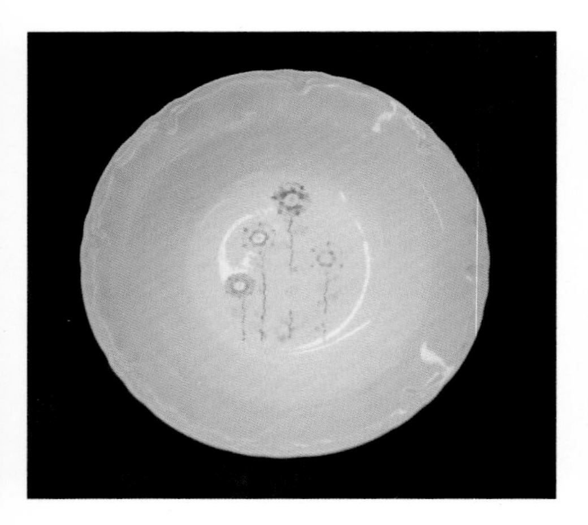

"PUNY POSIES," Scroll Edge, Bowl, $20.

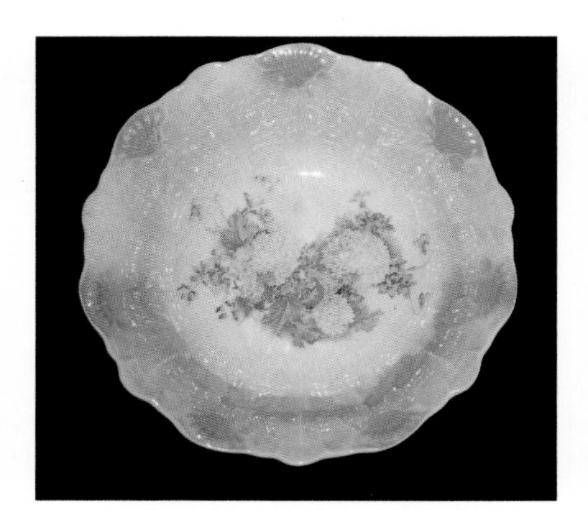

"LARRY'S LUSTRE," Lotus Leaf, Bowl, $25.

"ADAM'S FAMILY BOWL," Bowl, $25.

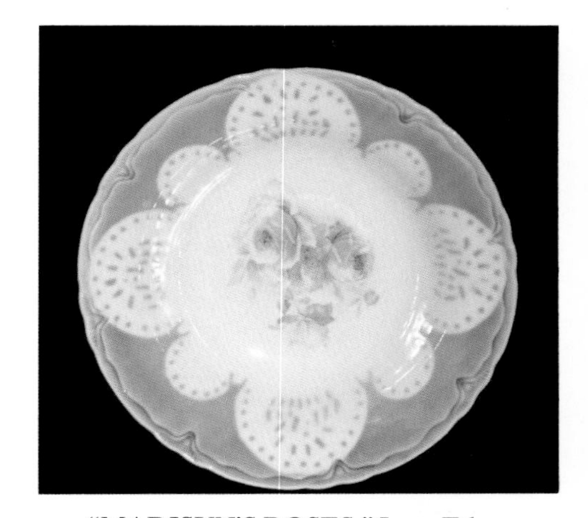

"MADISYN'S ROSES," Lace Edge, Bowl, $25.

EMMA (GREEN), Lace Edge, 9" Bowl, $35. This lace doily pattern is found in other colors.

'OLANDA," Decal Flower Bowl.

"CARMEN BETH," Bowl, $25. This and other granny bowls were sold for about $2 per dozen through catalogs. They were also given away by furniture stores and other merchants.

YELLOW TEA ROSE, Trellis, 10.5" Bowl, $35.

"DEEP ROSE," Squared Rib, Bowl, $25. Also found on Lace Edge and Scroll Edge shapes.

"COLLAGE," Lace Edge, Bowl, $25.

DOLLY, 9" Bowl, $25.

SPLASH, Bowl, $20. This pattern is also found in other colors with Scroll or Lace Edge.

CAPRICIOUS, Lace Edge, Bowl, $30.

OCTOBER BLUE, Square Rib, 9" Bowl, $25.

"STAMPED PANSIES," Scroll Edge, Bowl, $25.

GLENDA, Rib Edge, Bowl, $30.

BLUEBONNET, 10" Bowl, $45. This is a heavy bowl, very much like a mixing or salad bowl.

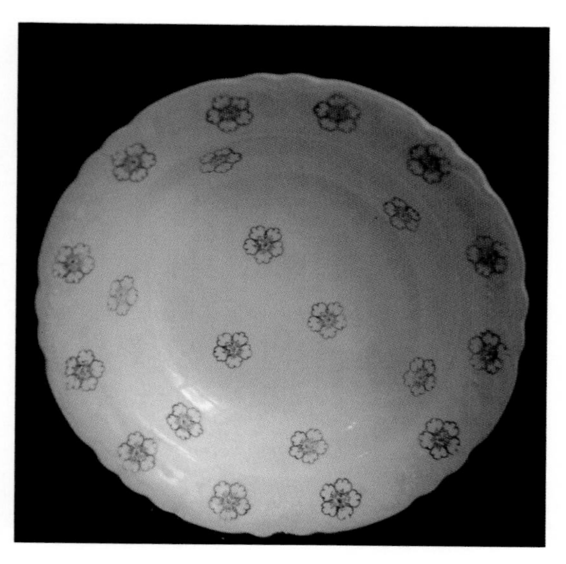

"DELFT DAISY," Scroll Edge, Bowl, $35.

"HERRINGBONE," 9" Bowl, $35. This bowl and many other Granny bowls were sold by Butler Brothers through their catalogs.

"BANANA SWIRL," Scroll Edge, 7" Bowl, $25.

DEBONAIR, Candlewick, 9" Bowl, $30.

"SIMPLY RIBBON," Astor, 9" Bowl, $25.

ROUNDELAY, Scroll Edge, 8" Bowl, $20.

"ENCORE (ROSE)," Granny Bowl, $20. Similar to MATILDA.

COTTON CANDY, Lotus Leaf, 10"Granny Bowl, $40.

"RUTH," Lace Edge, Bowl, $25.

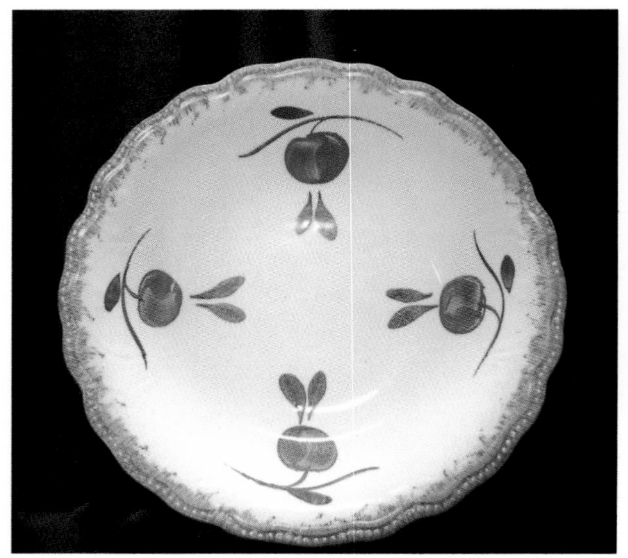

APPLE CIDER, Scroll Edge, 9" Bowl, $30. Very few Granny bowls with fruit patterns have been found.

SPINDRIFT, Lace Edge, Bowl, $25. DREAMY, Honeycomb, Bowl, $30.

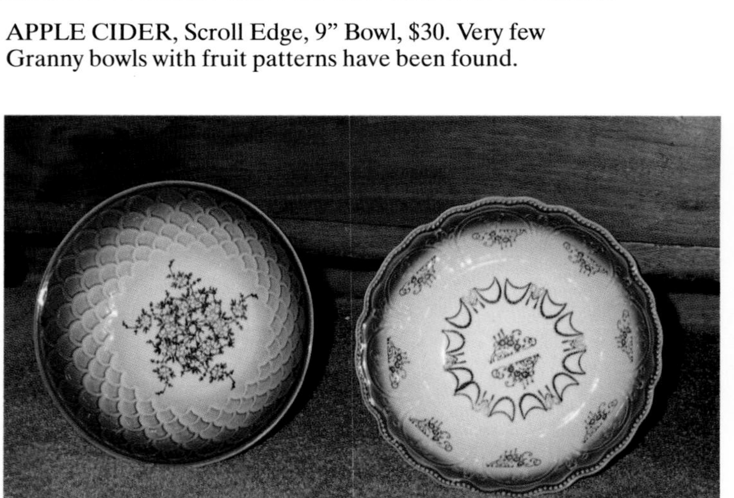

"BLUE POPPY," Honeycomb, Bowl, $35. S.P.I. backstamp. "ENCORE (BLUE)," Scroll Edge, Bowl, $25.

KELVIN, Bowl, $25.

ARDEN WEDDING, Honeycomb, Bowl, 30. This bowl with a blue or green border is BLUEBONNETS AND POSIES."

CHERISH, Lotus Leaf, 10" Bowl, $50. Also found on Squared Rib shape with aqua edge. CHERISH is also the name of a different pattern found on Colonial.

SPIDERWEB (GREEN), Salt and Pepper Shakers, $35.

"BURGUNDY SNOWFLAKE," Square Round Teapot, $125. Cake Lifter, $25. Also found in blue. Interestingly, sometimes it is stamped "coffee pot" in gold.

"BURGUNDY SNOW-FLAKE," Salt and Pepper Shakers, $30.

"DIVINITY," Gold Filigree Border, 9" Plate, $40. S.P.I. backstamp.

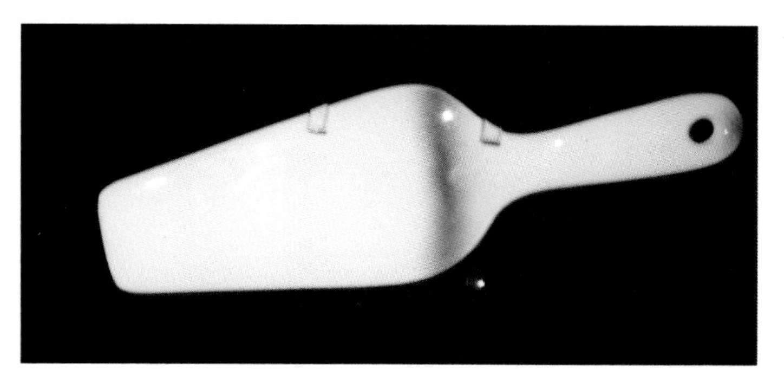

"FLOW," Cake Lifter, $25. This blue cake lifter may have been unfinished intentionally to go with several patterns or it may have missed its decal to become a BLUE SNOWFLAKE Cake Lifter. Measures 3.5" x 9".

"TURQUOISE CHAIN," Dinner Plate, $25.

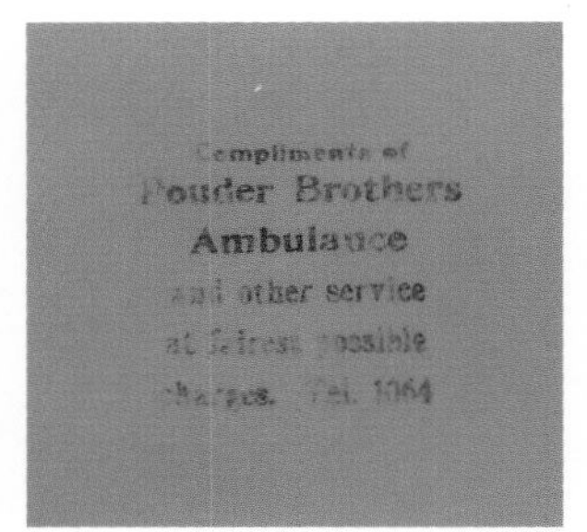

Back of "TURQUOISE CHAIN" Plate. Stamped: "Compliments of Pouder Brothers Ambulance and other service at fairest possible charges. Tel. 1064."

"CLASSIC GREEN," Lace Filigree Border, Dinner Plate, $30.

"LADIES IN WAITING," Plate, $100.

CHERUBS, Plate, $100.

Backstamp for CHERUBS and "LADIES IN WAITING" Plates.

ARCH THROUGH THE DESERT

CONVERTED TO THE CHRISTIAN FAITH

THE DEFENDERS OF THE CROSS

DANGEROUS INTRUDER

OFFERING OF PRAYER

ARRIVAL BEFORE JERUSALEM

ck of CRUSADE plates. Note the nchfield crown backstamp and Van ne Studio in script. The title of the ne depicted on each plate in the RUSADE set is found in the lower ht edge of its decal.

"ROMEO & JULIET," 9" Plate, $125.

Clinchfield crown backstamp on the back of "ROMEO & JULIET" plate. Also inscribed on the back in gold: "Van Dyne Studio." No information on this signature has been found.

TRIPLET, Monticello, 8" Plate, $25. This shape is sometimes referred to as waffle because of its edge treatment.

ARLENE, Trellis, Dinner Plate, $40. Cup and Saucer, $45. Painting inside the cup is very rare.

DAISY CHAIN, Square Handle, Open Sugars, $25 each.

"DAISY CHAIN VARIANT," Alice Pitcher, $125. Alice Pitchers are seldom marked.

"MONTY," Trellis, 12" Platter, $25.

BORDER PRINT, Square Rib, Cake Plate, $75.

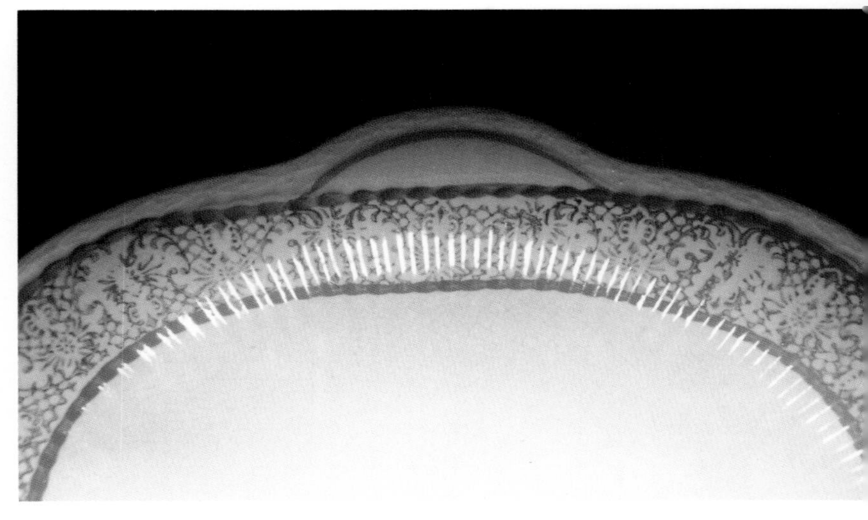

Close-up of Square Rib Edge on this very unusual piece.

Chapter Three
Floral and Foliage Patterns

Floral and foliage patterns have main-
ained their popularity among collectors.
ets in many patterns can be assembled at
somewhat reasonable price.

Handpainted Blue Ridge dishes on a handmade country quilt.

linchfield and Astor Shapes

SPRING SONG, Ball Teapot, $175.

FLAMBOYANT, Clinchfield, Tab Bowl, $25.

SHENANDOAH, Relish or Gravy Stand, $30. Usually found on Colonial shape. A very similar pattern on Candlewick is called STUCK UP.

IRENE, Astor, Snack Set, $40.

TEAL ROSANNA used in ad for this "Dishwashing Miracle." (Source Unknown)

CAROL ANN, Square Handle Creamer, $40.

"CALHOUN'S BLOSSOMS," 7" Plate, $15.

ALLISON, Clinchfield, Plate, $25.

"YALLISON," Clinchfield, Vegetable Bowl, $25. Pink-dotted brown center in each flower distinguishes this pattern from other daisy ones.

Backstamp on back of "YALLISON" Bowl. Note pattern number: 2667-U. The U denotes a variation of a previous pattern.

BUGABOO, Teapot, (with lid), $125.

BOUQUET, Ball Teapot, $150.

HOPSCOTCH, Platter, $35.

"MARION," Clinchfield, 8" Plate, $25.

.HOBOTH, Creamer, $25.

e other side of the REHOBOTH
eamer.

NONSENSE, Astor, Fruit Bowl, $5.

JESSICA, Astor, 6" Plate, $8. Also found on Colonial shape.

WILDWOOD FLOWER, Clinchfield, 8" plate, $30.

TIGER EYE, Dinner Plate, $20.

PHLOX, Astor, Dinner Plate, $25.

RECOLLECTION, Astor, Dinner Plate, $20. This pattern on Colonial shape is ERWIN SPRING.

"DWARF TULIPS," Astor, Dinner Plate, $25. This pattern is very similar to #7 of the PHYLLIS salad set.

BLUECURLS, 9" Plate, $30.

MAYTIME, 7" Plate, $40. Usually found on Colonial shape.

"SIDE BY SIDE," Plate, $20.

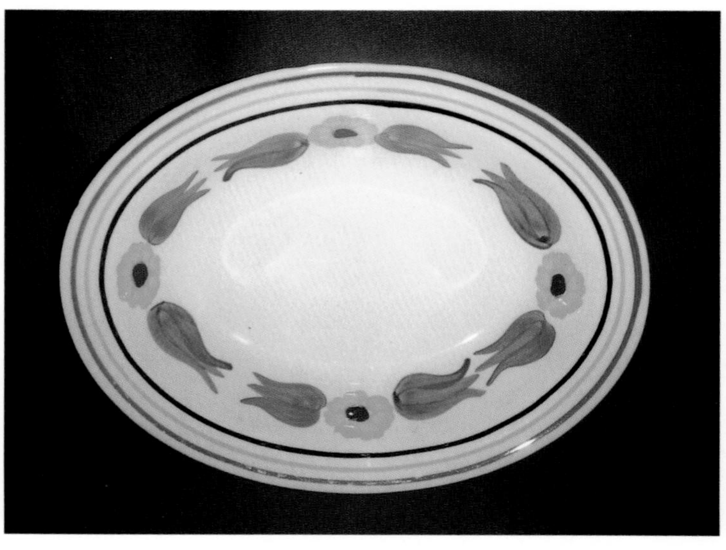

CHAMBLEE, Oval, Vegetable Bowl, $35.

UTHERN ROSE in The February, 1948 issue of
TTER HOMES AND GARDENS.

NASSAU, Astor, Dinner Plate, $25. Gravy Boat, $30.
Gravy boat fits nicely on the bread and butter plate.

BEVERLY, Astor, Dinner Plate, $20.

Back of BEVERLY plate with name and backstamp.

ENSY," Astor, Cup and Saucer, $25.

CHICHORY, Covered Toast, $150.

"GERT," Dinner Plate, $25.

Lots of Blue Ridge tucked safely away in these lawyer's shelves. Note SHANNON and PEONY on the bottom shelf.

"POP'S RIBBON ROSE," Big Cup and Saucer, $75. This cup is only 3".

"POP," painted on the other side of this RIBBON ROSE Big Cup.

"KATE'S ROSE GARDEN," 17" Platter. Reportedly, this platter is a painter's piece, one which the painter kept.

ENGLISH GARDEN, Saucer, $5.

BRECKENRIDGE, Partial Breakfast Set.

DEMOREST, Big Cup and Saucer, $40. Big saucers are 7.5" and deeply fluted. Matching cups are usually 4". These sets are also called "Jumbo."

"THE PRIZE," Flat Soup Bowl, $30. Marked #3623 on back. Similar to FAIRY TALE.

ITTER, ROSEANNA, Toast Covers, $50 each. These fit uare plates or 8" round plates and are part of a breakfast set.

t of water glasses in ROSE OF SHARON pattern.

EGLANTINE, Clinchfield, Saucer $8.

YELLOW RIBBON, Platter, $30.

ROSE OF SHARON, Dinnerware. Usually found only on accessory pieces.

BABY DOLL, Square Round Teapot, $200.

SYDNEY'S HOPE, Flat Soup Bowl, $30.

WILDWOOD, Creamer, $25.

FLOWERY BRANCH, Astor, Dessert Plate, $15. Full pattern has an additional flower.

"DEANNA," Saucer, $5.

TULIP CORSAGE and golden mums offer a beautiful table for any occasion. (Source Unknown)

"TERRI ANN," Flat Soup Bowls, $20 each.

"KRIS LEIGH," Cake Plate, $45. Notation on back, "1213-US."

BLARNEY, Astor, Covered Sugar, $25. A similar pattern on Candlewick is WAVERLY.

"VALLEY BLOSSOM," Coaster, $45. Backstamp includes the name of the pattern, as well as Blue Ridge Mountains, HAND ART.

'ROSIE," Vegetable Bowl, $25.

"AROUND ROSE MARIE," Covered Toast Underplate, $45. This is an unusual treatment of the ROSE MARIE pattern on the Astor shape.

BBY ROSE, Dinner Plate, $25.

BLUE FLOWER, Big Cup and Saucer, $100.

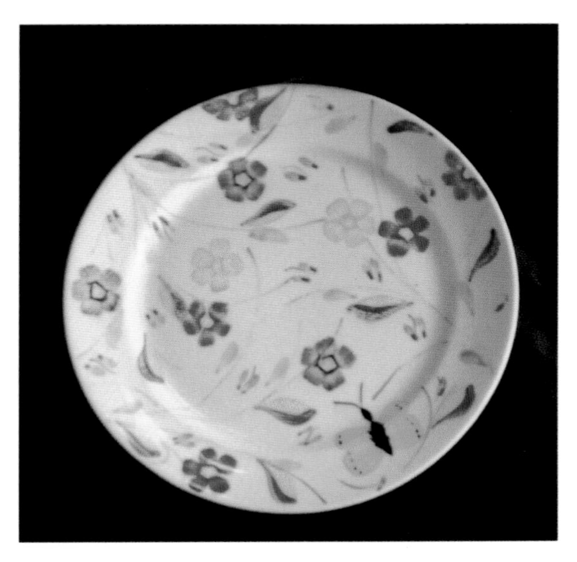

SUE-LYNN, Astor, Dinner Plate, $40.

SSE, Dinner Plate, $20.

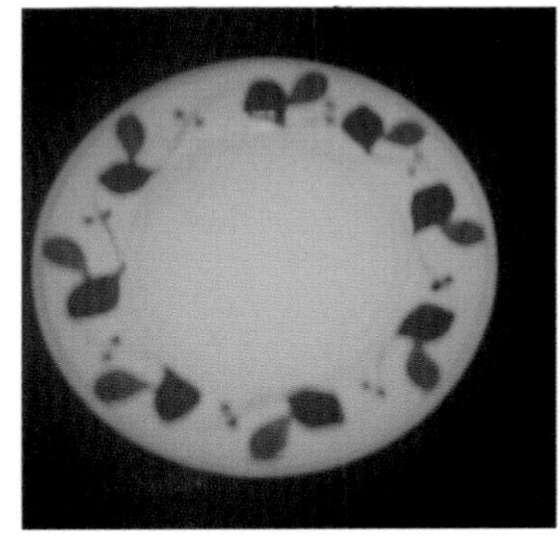

FANCY FREE, Clinchfield, Dinner Plate, $25.

EVERLASTING, Clinchfield, Dinner Plate, $25.

"SHEILA," Clinchfield, Bread and Butter Plate, $15.

"MAUDE VARIANT," Flat Soup Bowl, $15.

TRIBUTE, Astor, Big Cup and Saucer, $75.

SUMMER SUN, Saucer, $4.

DAISY CHAIN, Square Round Teapot, $95 with lid.

HEX SIGN, Clinchfield, Fruit Bowl, $6.

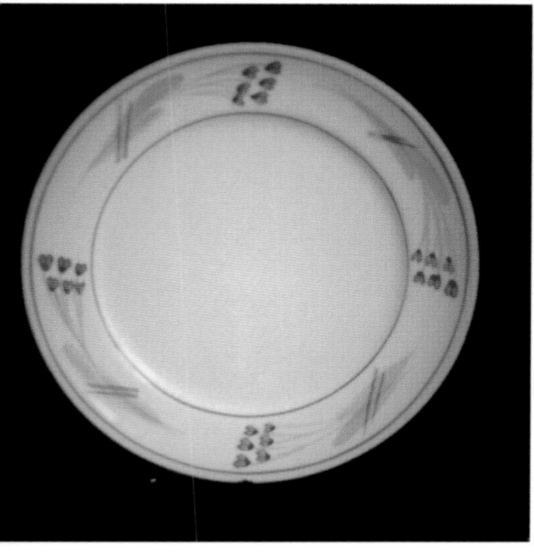

ALENTINE ASTERS, Astor, Bread and
utter Plate, $8.

BRISTOL BOUQUET, Square Handle
Creamer, $50.

YELLOW PLUME, Astor, Covered Toast, $100.

TANSY, Demi Cup and Saucer, $40. TANSY, Big Cup and Saucer, $70.

Colonial Shape

t of FUCHSIA.

"YELLOW RAMBLER," Dinner Plate, $25.

MADRIGAL, Saucer, $5. The demitasse pieces in this pattern are called ROSY FUTURE.

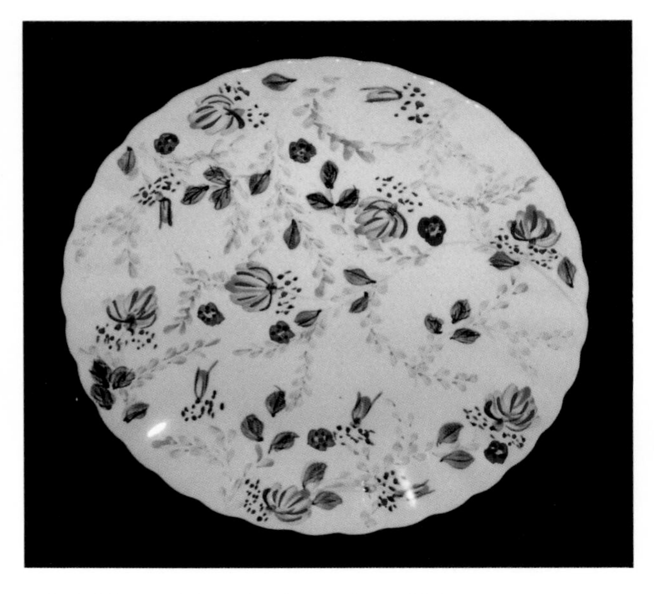

"JEAN'S DELIGHT," Dinner Plate, $30.

BRIAR PATCH, Dinner Plate, $30.

CHICKORY, Colonial, Teapot, $125.

CHRISTINE, Saucer, $7.

Flowery plates on flowery wallpaper.

REMEMBRANCE, Dinner Plate, $30. This plate on Astor is called TENNESSEE WALTZ.

NNALEE, Dinner Plate, $25.

ROAN MOUNTAIN ROSE, Oval, Vegetable Bowl, $30.

GEORGIA BELLE,
Two-Tier Tidbit, $30.

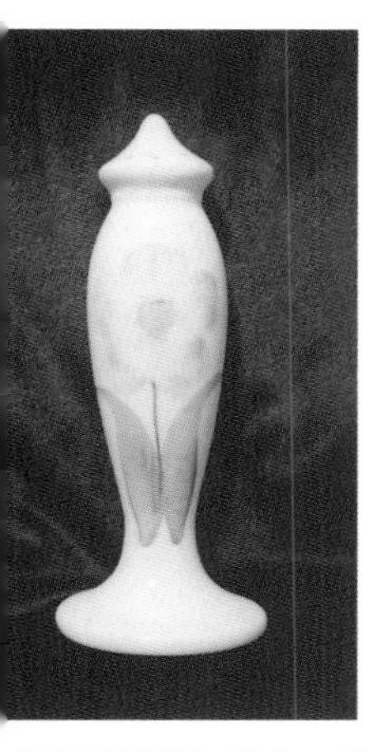

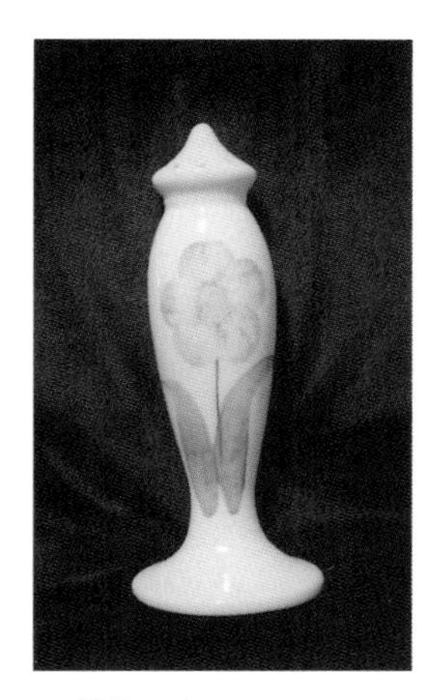

ADIANCE, China Shakers, $75, pair.

MOUNTAIN BELLS, Dinner Plate, $20.

Southern Potteries, Inc. label on the back of MOUNTAIN BELLS plate. The label indicates that this is Dec. No. 53898.

"FLOWER SONG," Luncheon Plate, $25.

KELCI, PRIM POSIES, Dinner Plate, $20. This pattern was named "PRIM POSIES" by Winnie Keillor in 1983.

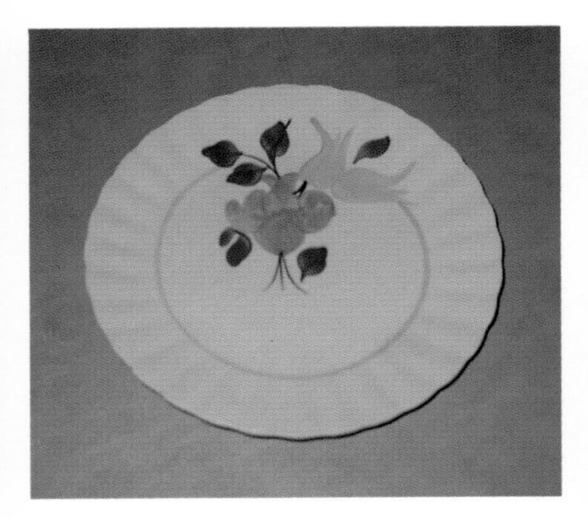

TULIP, Plate, $20. This pattern was previously known as JUNE BRIDE.

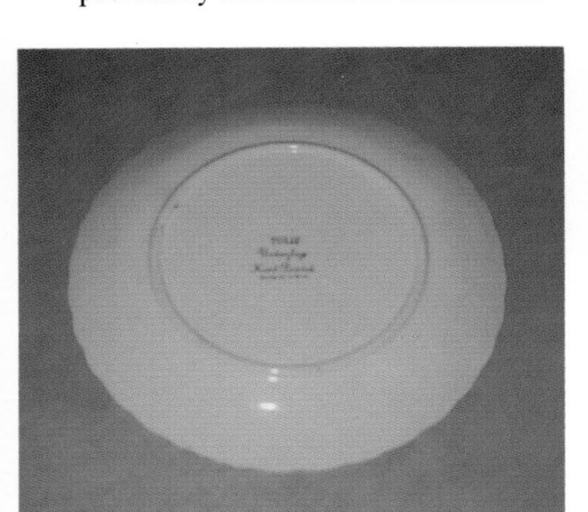

Backstamp showing "TULIP" name.

SNIPPET, Ovide, Coffee Pot, $250.

WRINKLED ROSE (YELLOW EDGE), Ovide, Coffee Pot, $150.

VIOLETS FOR JEAN, Colonial, Teapot, $150.

SNIPPET, Colonial, Teapot, $200.

SNIPPET, Covered Toast Cover. Note the steam hole to keep toast crisp.

VIOLET," Snack Set, $55. Backstamped: Blue
Ridge Mountains Hand Art.

VIOLET CIRCLE, 4" Tea Tile, $50.

An elegant table set with FLOWER
BOWL. (Source Unknown)

NGELINA is a "friendly" pattern. SPRING BEAUTY is a variant.

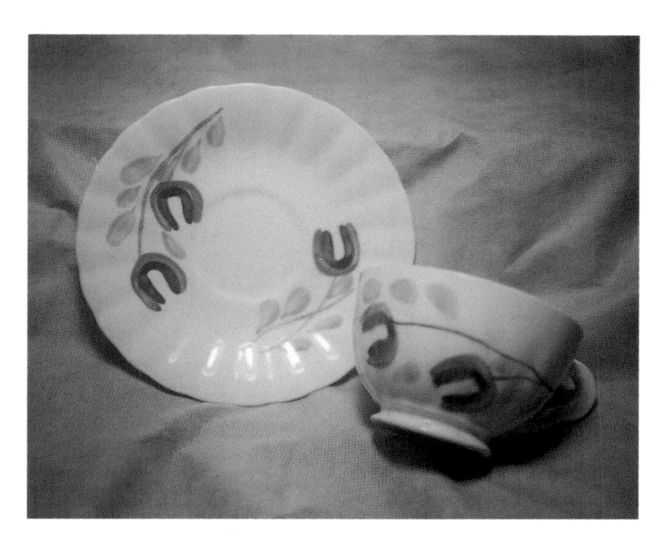

CROCUS, Cup and Saucer, $20. CROCUS also refers to
a totally different pattern.

LAURA, Colonial, Teapot, $150.

LAURA, Dinner Plate, Cup, and Saucer. Back of plate marked pattern
2754-U. Very difficult pattern to paint.

CARLILE, Barrel Shaker,
$40 pair.

ROCK ROSE ("VALLEY BLOSSOM,") Plate, $20.
Backstamp shows "VALLEY BLOSSOM" as name of pattern.

VIOLA, Colonial, Coffee Pot, $175.

"ROCK ROSE VARIANT," Cake Plate, $45. 6" Plates, $10 each. Cake Lifter, $30.

CAROL'S CORSAGE, Dinner Plate, $25.

STARDANCER, 10" Plate, $75.

"TARA'S ROSE," Two-Tier Tidbit, $45.

"KAYE," Dinner Plate, $20. "BLUE EYES," Platter, $20.

OLKA DOT, Snack Set, $45. One of the few atterns with flowers painted inside the cup.

GYPSY DANCER, Colonial, Teapot, $175. Lid is for WRINKLED ROSE, Lid only, $10.

CORDELE, Dinner Plate, $20.

EDGEMONT, Colonial, Teapot, $150.

Left to right) Practice, or unfinished, plate for BRIDESMAID. BRIDESMAID Bread and Butter Plate, $8.

ASTER BLOSSOM, Colonial, Square Plate, $25.

CHAMPAGNE PINKS, Ovide, Coffee Pot, $150.

RUTH ANNA, Teapot, $200.

CHARMER, Tab Bowl, $15. Look for a red flower on the full pattern.

CINNABAR, Bread and Butter Plate, $7.

BREATH OF SPRING, Saucer, $5. Also found without green edge. Whole pattern also has a yellow flower.

SHEREE, Bread and Butter Plate, $5.

FLOUNCE, Bread and Butter Plate, $5.

PENNY SERENADE, Bread and Butter Plate, $5.

TERESA, Sugar, $20, with lid. Full pattern has a five-petal yellow flower and a three-petal rust bud.

LAVENDER IRIS plate on the right has not been painted in the center. These "human error" pieces are very appealing to collectors.

ıis frosted glass refrigerator jar comple-
ents several Blue Ridge patterns. The single
insettia and the yellow mums resemble the
tterns of the GARDEN FLOWERS Salad
t. The manufacturer is unknown.

YELLOW MUMS (GARDEN FLOWERS), 9" Plate, $30. This pattern on the Clinchfield shape is called JANA.

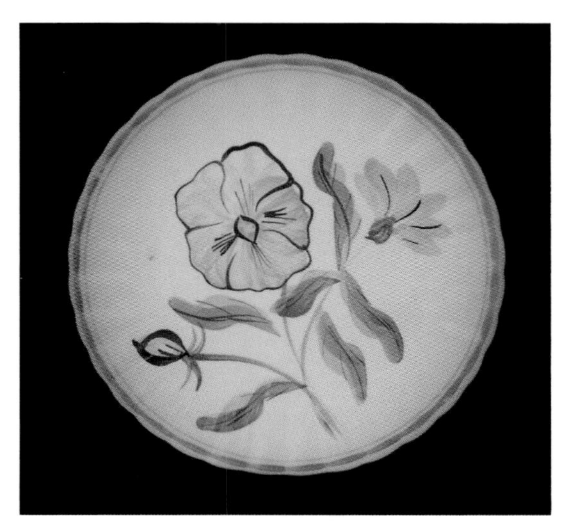

YELLOW PANSY, (GARDEN FLOWERS), Luncheon Plate, $30.

PURPLE MAJESTY, Salad Plate, $25. Named by Winnie Keillor. An almost identical pattern with two flowers is #4 in the FONDEVILLE FLEURS Salad Set. It has a Fondeville, New York, backstamp.

HOLLYHOCK, (GARDEN FLOWERS), Salad Plate, $30.

"PINK MORNING GLORY," Ashtray, $25.

Beautiful shelf with garden flowers. Note PURPLE CROWN tapered vase on left end of middle shelf.

(Left to right) PURITAN, Bread and Butter Plate, $5. "SONDRA," Bread and Butter Plate, $5.

"WILD PETUNIA," Salad Plate, $25.

JOSEY'S POSIES, Square Plate, $25. This pattern was made for PV (Pleasant Village).

"WINTER PANSIES," Dinner Plate, $30.

EXUBERANT, 9" Bowl, $30. Also found on Colonial shape.

"WHIRLY-WHIRLY," Bread and Butter Plate, $8. This pattern is similar to WHIRLIGIG.

"MUSEUM PIECE," 9" Plate, $25.

TIGER LILY on 1950s tablecloth.

BLACKBERRY LILY, Ovide, Coffee Pot, $125.

FLOWER RING, Ovide, Coffee Pot, $200.

WHIRLIGIG, Cup, $25. The price for this cup without the pattern number is $10.

Pitchers and glasses to match the NORMA pattern. This set was purchased, in the original box, in the 1970s.

NORMA, Service for four in the original shipping box.

GRESHAM, Platter, $30.

RHAPSODY, Ovide, Coffee Pot, $150.

PANDORA, Platter, $30.

ˌAREPTA, Salad set, $150. A salad set consists of salad bowl, ˌderplate, fork, and spoon.

ERWIN, 12" Platter, $30.

ˌNCHANTMENT, Fruit Bowl, $5. ˌll pattern also has a bud.

"EDDIE," Bread and Butter Plate, $7. Full pattern also has a yellow flower.

ADALYN, Dinner Plate, $20.

OLGA, Saucer, $5.

"TOMORROW," Bread and Butter Plate, $5. LOUISA, Bread and Butter Plate, $5.

NOCTURNE (RED), Wall Lamp made with Plate, Cup, and Saucer.

BECKY, Colonial, Teapot, $150.

FLIRT, Oversized Open Sugar and Creamer, $45 set. FLIRT is more often found on Piecrust.

WILD STRAWBERRY, RUTH ANNA, MIRROR IMAGE, WIND-FLOWER Egg Cups, $40 each. These originally sold for $1 each.

KATE, 10" Plate, $30.

"RED GAIETY," Advertisement in HOUSE BEAUTIFUL, July, 1948.

WILD ROSE, Oval, Vegetable Bowl, $22.

"PATTY," 6" Round Tile, $35.

"OUR LISA," 9" Bowl, $25. Here again, the decorator "fits" the pattern into the shape.

BRISTOL BOUQUET, Dinner Plate, $25. This pattern on Astor has a red line around the edge.

Neatly displayed sets of (top to bottom) CYCLAMEN, (RED) WILD IRISH ROSE, JUNE BOUQUET, CHRYSANTHEMUM, and LAURA.

SUNDAY BEST, Colonial, Teapot, $150.

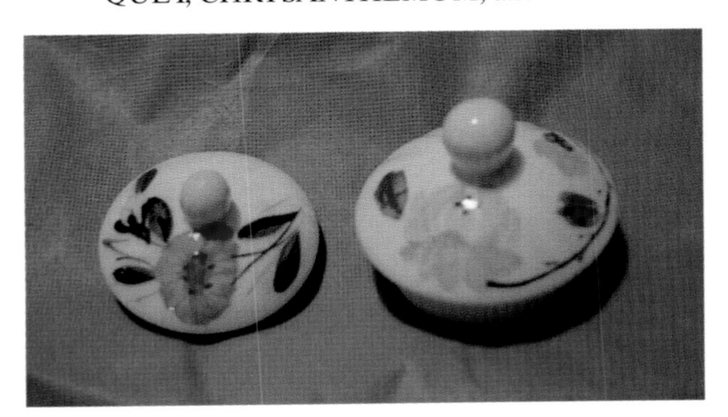

(Left to right) KING'S RANSOM, Sugar Lid, $5. KAREN, Colonial, Teapot lid, $15. The sugar lid will also fit the Ovide Coffee Pot and is often used to replace broken lids.

MEYLINDA, 9" Bowl, $25.

KAREN, Colonial, Teapot, $175.

PEONY, Ovide, Coffee Pot, $200, with correct lid.

PEONY, Colonial, Teapot, $175 (with lid.)

MEAGAN, Dinner Plate, $15

"RED DAISY," Dinner Plate, $30.

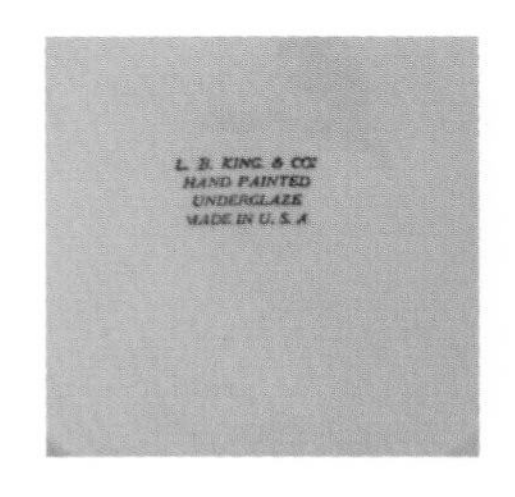

Backstamp on "RED DAISY" plate: I.B. KING & CO., HAND PAINTED, UNDERGLAZE, MADE IN U.S.A. Sometimes this jobber's backstamp will simply read: KING'S, HANDPAINTED, UNDERGLAZE.

RIDGE ROSE pieces shown here are from the estate of the late Marion La Bar Searfass. RIDGE ROSE, Teapot, $150.

RIDGE ROSE, Cake Set with Cake Lifter, $275.

DGE ROSE, Salad Set, $150.

PANSY TRIO, Dinner Plate, $35.

"PRISSY," Luncheon Plate, $15.

LEA MARIE, Dinner Plate, $20.

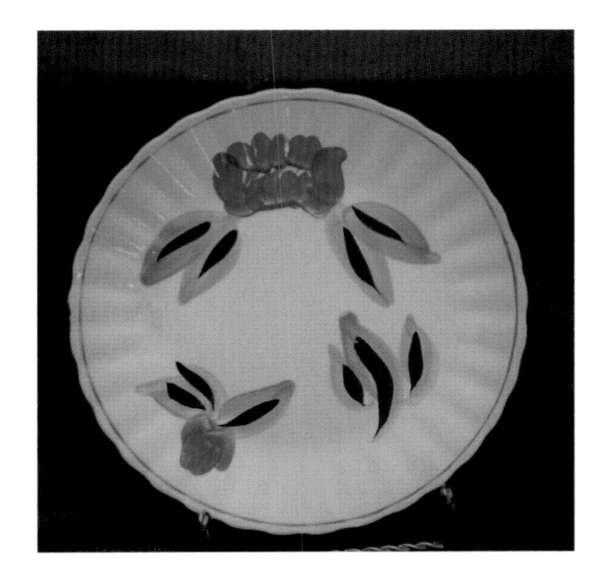

KENNESAW, 9" Plate, $25.

PETUNIA, (LAUREL WREATH), Ovide, Coffee Pot, $200.

ARABELLA, Dinner Plate, $15.

MEADOW BEAUTY, Salad Plate, $15.

(Left to right) VARIETY Gravy Boat, MULTICOLOR TULIP TRIO Plate, and TULIP ROW Teapot. The colors are reversed on the teapot.

CHATHAM, Bread and Butter Plates, (left) $5, (right) $35. The plate on the right was glazed before the full pattern was painted, making it highly collectible.

PETITE FLOWER SET, (Top to bottom, left to right) "PETITE STARFLOWER," "PETITE CLOVER," "PETITE RED TULIP," and "PETITE YELLOW TULIP," $30 each.

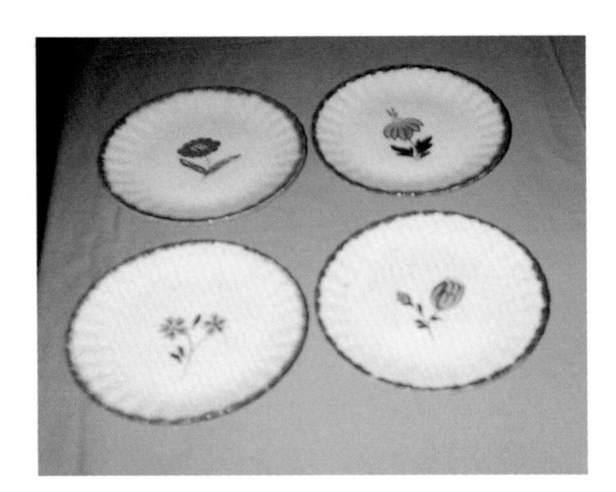

Above: PETITE FLOWER SET, (Top to bottom, left to right) "PETITE RED DAISY," "PETITE SNOWDROP," "PETITE VIOLETS," and "PETITE BLUE TULIPS," $30 each.

Right: TWO OF A KIND on a 1950s tablecloth.

.LEENA, Dinner Plate, $15.

"TWO OF A KIND VARIANT," Dinner Plate, $20.

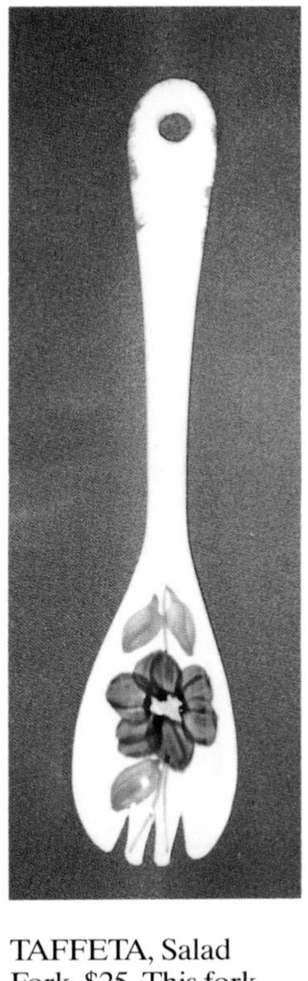

TAFFETA, Salad Fork, $25. This fork and accompanying spoon could match other patterns.

OPKNOT, Dinner Plate, $15.

JUDITH, Dinner Plate, $20.

ASHLAND, Dinner Plate, $25.

COVINGTON, Tidbit. Note the ornate handle.

RED HILL, Colonial, Teapot, (with lid) $150.

SHERRY, Dinner Plate, $15.

IRISH MARY, 10" Salad Serving Bowl, $50.

Colonial Faience backstamp on bottom of IRISH MARY Salad Bowl. Note the Candlewick edge found around the rim of salad bowls.

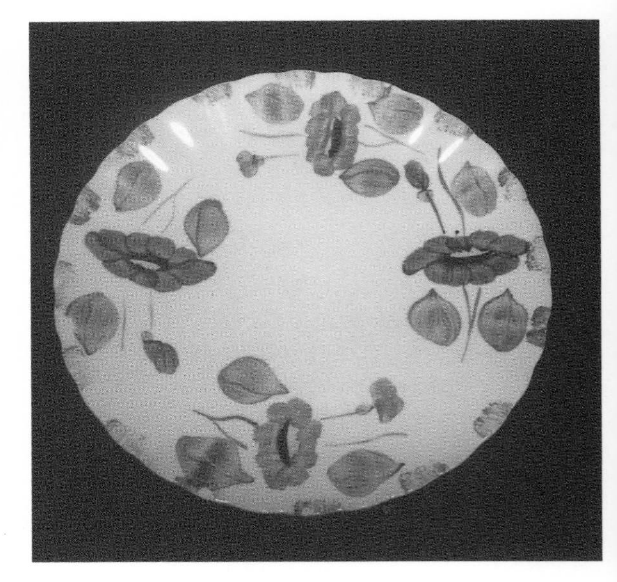

FAIRFIELD, Dinner Plate, $20.

"ANNELIESE," Dinner Plate, $20.

GARLAND, 7" Square Plate, $30.

UPSTART, Well Platter, $25. Usually found on Skyline.

AMANDA, 9" Plate, $20.

'ACCATO, 6" Plate, $5. This
ttern is also called RED STAR.

JUNE BOUQUET, Colonial, Teapot, $125.

Glasses in two sizes to go with SUN BOUQUET and other similar patterns.

ELD DAISY, Colonial, Teapot, $125.

"SANKIE," Dinner Plate, $18.

MOUNTAIN DAISY, Bread and Butter Plate, $5.
Backstamp showing name of pattern.

"DOUBLE DAZZLE," Dinner Plate, $22.
Bread and Butter Plate, $8.

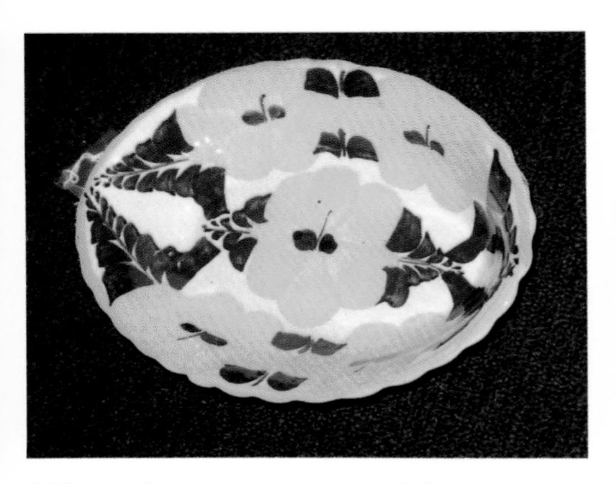

"INEZ," Oval, Vegetable Bowl, $20.

LACE-LEAF COREOPSIS, Dinner Plate, $15.

TEXAS ROSE, Creamer, $25.

MIRROR, MIRROR, Colonial, Teapot, $150. Lid is for SUNFLOWER.

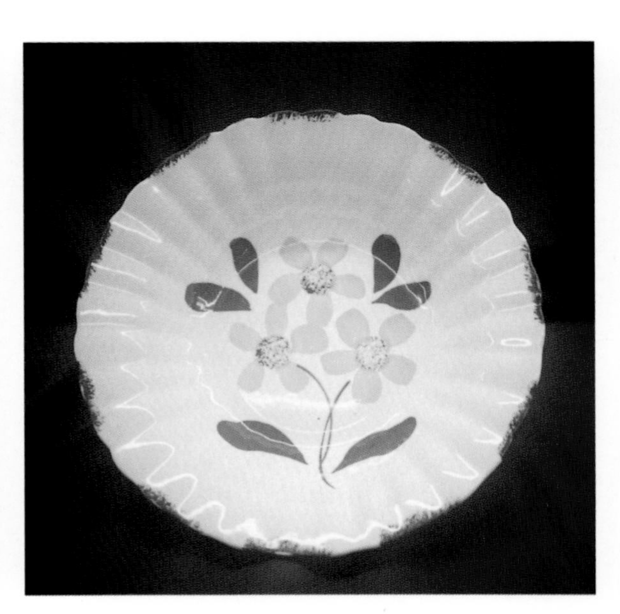

CHEERFUL, 8" Bowl, $20.

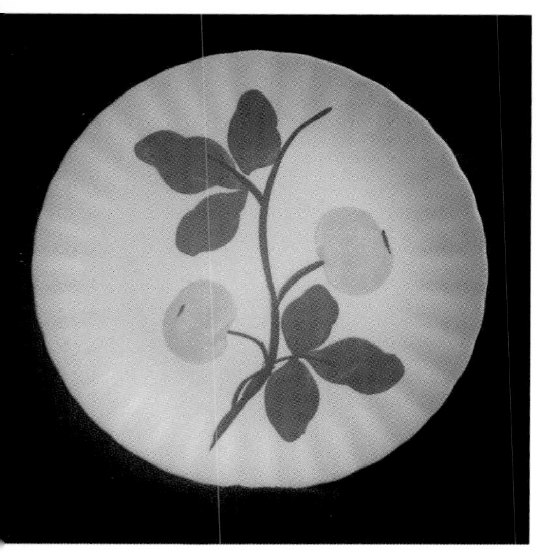

OSMIC, Luncheon Plate, $15.

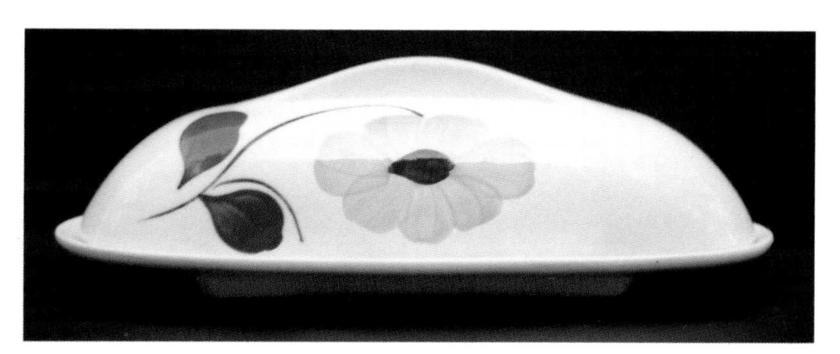

RIDGE DAISY, Butter Dish, $75. This style of butter dish was used with most shapes.

Glass Set also matches RIDGE DAISY.

HALF & HALF, Dinner Plate, $25.

IDGE DAISY, Matching Pitcher.

RUGOSA helps to sell Sanka, a 97% caffeine-free coffee. (LIFE Magazine, November 7, 1944.)

MOUNTAIN GLORY (YELLOW NOCTURNE) and MOUNTAIN LAUREL (PETUNIA/LAUREL WREATH) advertised by Marshall Field & Company in the Chicago Sunday Tribune on August 30, 1942. This 24- piece service for four sold for $5.50!

MOUNTAIN GLORY, Big Cup Saucer, $35.

Backstamp on MOUNTAIN GLORY Big Saucer, indicating name of pattern. Also stamped: Blue Ridge Mountains, Hand Art.

YELLOW NOCTURNE, Salad Fork, $25. Tines on this type salad fork were hand cut from the salad spoon mold.

VANITY FAIR, Dinner Plate, $15.

SHOO FLY, Traditional Colonial, Creamer, $20. Creamer and Open Sugar Set, $25. These pieces are interchangeable.

"SABRE," Dinner Plate, $15.

YELLOW NOCTURNE, Colonial, Teapot, $125.
YELLOW NOCTURNE, Cake Plate, $45.

COWSLIP, 12" Platter, $25. COWSLIP and FASCINATION are almost identical patterns.

VERONA, Platter, $25.

"JONATHAN'S MEMORY," Platter, $30.

HIMSEY, Saucer, $4.

BLUE FLOWER, Dinner Plate, $20.

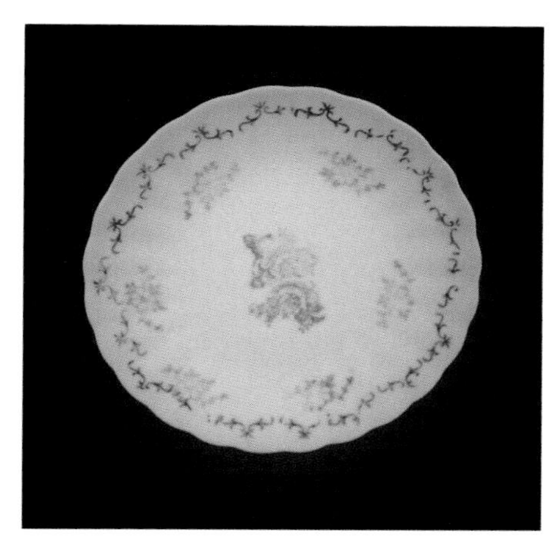

"SALEM," Bread and Butter, $5.

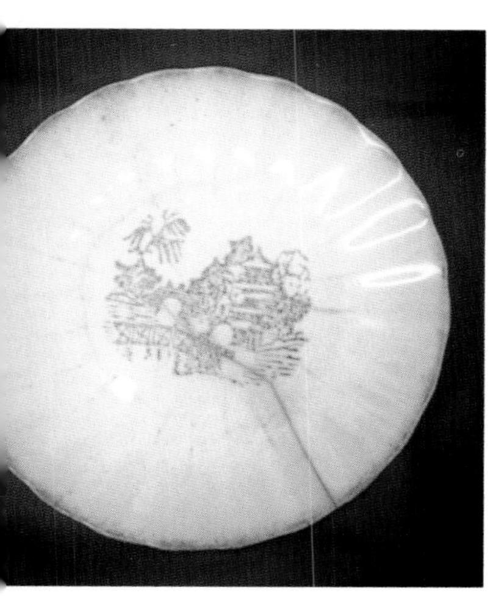

REEN WILLOW," Dinner Plate, $25.

MISHA, Dinner Plate, $20.

"OUCH," Dinner Plate, $25. This plate has Temp 717A on the back.

WITCHERY, Saucer, $4.

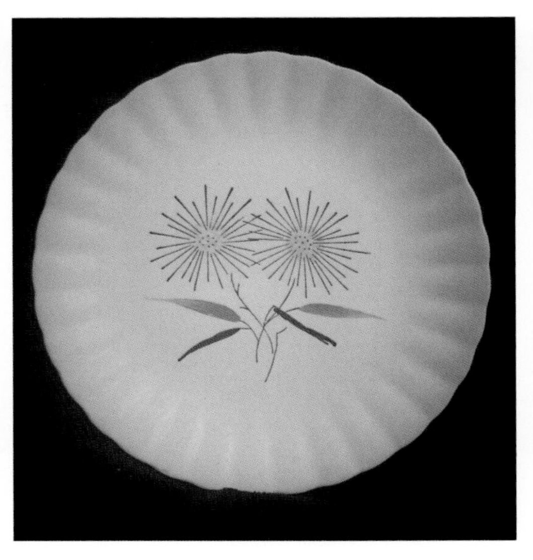

REFLECTION, Dinner Plate, $15. Also found on Astor shape.

"SQUIGGLE," Bread and Butter Plate, $5.

FOX GRAPE, Brass Candle Holder.

GRANDMOTHER'S GARDEN, 11" Plate. Signed Edna E. Kerr.

SUMMERTIME, Dinner Plate, $40. This pattern is usually found only on accessory pieces.

NOVE ROSE, Platter, $125. Each decorator seemed to add a brush stroke here or there!

ALTZ TIME, Ovide, Coffee Pot, $125.

ELEGANCE, 14" Platter, $85.

ROMANCE, Salad Bowl, $75.

ROSE MARIE, Cup and Saucer, $65. One of the few cups with decoration inside.

ccessory Pieces

ROSE MARIE, Chick Pitcher, China, $225.

ROSE MARIE, Chocolate Pot Lamp, $225.

"MIDNIGHT," Chocolate Pot Set, $1,200. ROMANCE, Chocolate Pot Set, $1,100.

EASTER PARADE, Chocolate Pot, $225.

"MIDNIGHT," Chocolate Set Tray, $750. Back of the "MIDNIGHT" Chocolate Set Tray is marked Tray #2. This gently scalloped tray is approximately 15" by 9", just large enough to comfortably hold the chocolate pot, pedestal creamer, and pedestal sugar.

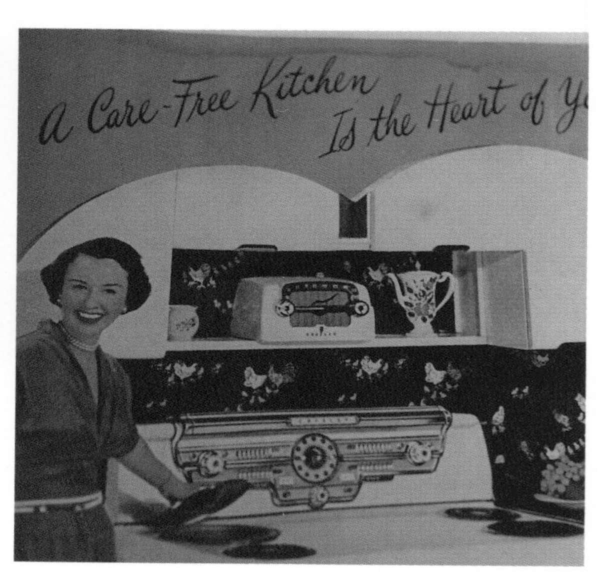

Geneva kitchen ad with ROSE MARIE Chocolate Pot, SATURDAY EVENING POST, 1953.

CALICO, Chocolate Pot, $200.

ROSE OF SHARON, Chocolate Pot Set, $1,000. Shakers, $100.

OSEMARIE, Pedestal Sugar, Creamer, nocolate Pot Lid, and Bottom of Chocolate ot. Inscribed: June 12, 1944.

ELEGANCE, Milady Pitcher, $300.

"WANDA," Sally Pitcher, $125.

LL COLORS, Flared Creamer, $65.

MELODY, Helen Pitcher, $150.

Inside view of "WANDA" Sally Pitcher.

"MAGGIE LEE," Flared Sugar, $75.

BLUE IRIS, 7" Spiral Pitcher, $125. Note slight variation in painting of BLUE IRIS due to size differences.

CONASUAGA, Virginia Pitcher, $175.

CALICO, Clara Pitcher, $125.

UE IRIS, Mini Spiral tcher, 4.5", $125.

Other side of BLUE IRIS Mini Spiral Pitcher, $125.

"WHISPER," Virginia Pitcher, $120.

ERWIN ROSE, Virginia Pitcher, $125.

"SISTERS," 4" Mini Virginia Pitcher, $150.

"MINNIE," Alice Antique Pitcher, $200.

"MOTLEY BLUE," Martha Pitcher, $100.

MELODY, Jane Pitcher, $125.

ROSE OF SHARON, Antique Pitcher, 5", $125.

GOOD HOUSEKEEPING ROSE, Mini Ball Teapot, $325.

aborate collection of ANNIVERSARY SONG. Note Milady Pitcher and Ruffle p Vase among these pieces.

ROSE OF SHARON, Fine Panel, Teapot, $200.

"IN YOUR DREAMS," Teapot. Reportedly, one of two painted by Lola Johnson Bailey, which recently sold for $875.

OUNTAIN NOSEGAY, Mini Square Round Teapot, $350.

BORDER PRINT, Square Round Coffee Pot, $125. Several pots have been found marked Coffee Pot on the bottom.

Good Housekeeping pieces are exceptional and rare, especially "GOOD HOUSEKEEPING VIOLETS."

"GOOD HOUSEKEEPING VIOLETS," Teapot, $350. Mini Ball teapot in this pattern is called GORDON VIOLETS.

ALLISON, Square Round Teapot, $125.

"CHRISTY," Good Housekeeping, Teapot, $150.

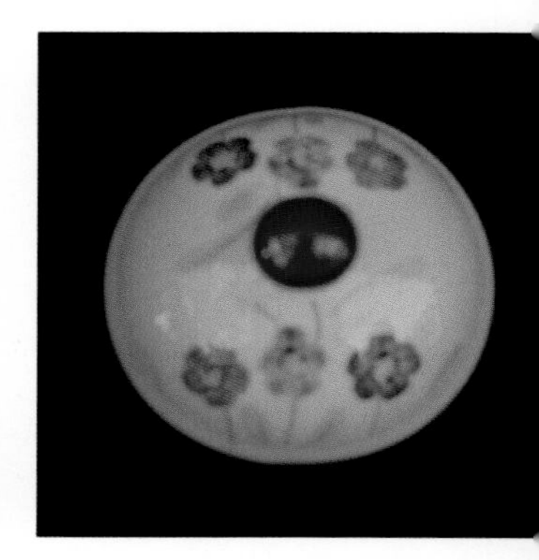

Lid for YELLOW ROSE Teapot, $25.

"CHRISTY," Good Housekeeping, Creamer, $50.

STEPHANIE, Handled Lamp, $175.

Bud Top Shakers, "GOOD HOUSEKEEPING VIOLETS," $150. Look for these in Blossom Top.

ILBERTINE, Handled Vase, 75. Stamped: Genuine Porce-in, Vanity Fair.

MELODY, Teapot Lamp, $150.

JUNE BOUQUET, Teapot Lamp, $125.

OGTOOTH VIOLET, Handled Vase, 50. Vase is stamped: Southern Potteries, nity Fair.

LARGO, American Home Lamp, $125. Hole near base for cord.

Slight differences in height and vase opening are a result of different molds.

HIBISCUS, Bulbous Vase, $125. Note how the flower is not painted over the embossing, but rather to the side of it. The embossing fits the HAMPTON vase shown in *Blue Ridge China Today*.

ROSE OF SHARON, Bulbous Vase, $150. If you look closely, you can see the embossed flower on this mold.

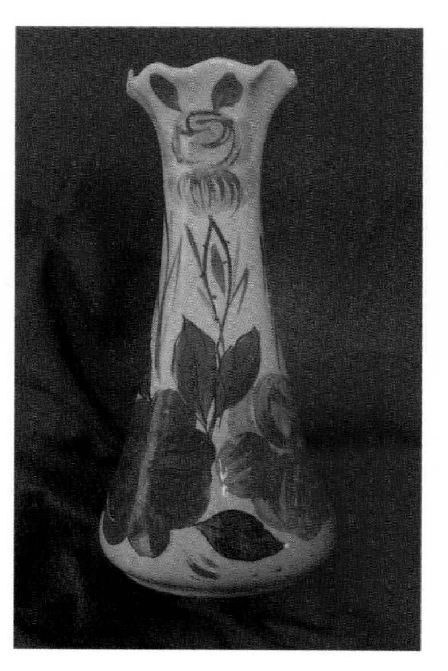

DELPHINE, Ruffle Top Vase, $125.

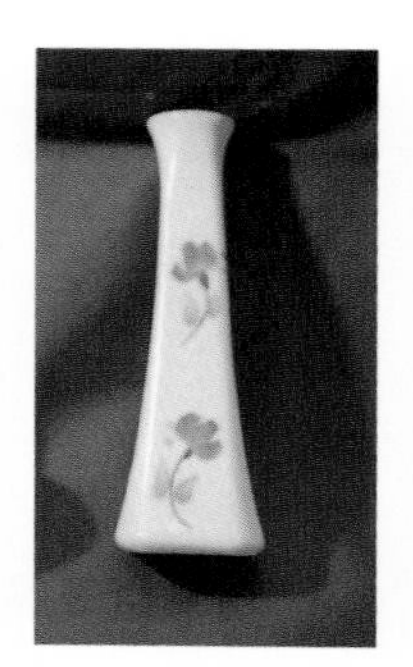

FLIRT, Mini Vase, $150. These dainty treasures are hard to find.

RIDGE DAISY, Bud Vase, $150.

CHINTZ, Tapered Vase, $120.

Display case of beautiful accessory pieces and demitasse sets. Note "GOLDEN GIRL," Art Nouveau, Flower Frog.

MOOD INDIGO, two different 7.25" Tapered Vases. Both have the Blue Ridge China backstamp.

"MARTHA'S EGGS," Egg Plate.

Both sides of a lovely Boot Vase with the RECOLLECTION, Astor, and ERWIN SPRING, Colonial patterns.

IXIE, Flat Shell Bon-Bon, $125. PIXIE ornament purchased
: the Apple Festival.

CHARM HOUSE pieces are a favorite of collectors. Pictured here are a Ramikin, $200, Sugar and Creamer, $150 (pair), Pitcher, $250, and Server, $200.

ARDEN LANE, Martha Snack
ay, $200.

ROSE OF SHARON, Deep Shell, $125.

NOVE ROSE, Maple Leaf Relish, $75, shown with painting of the same piece.

CONASAUGA, Heart Relish, $100.

CHARM HOUSE, Relish, $200. This 6.5" by 5" shaped piece is unique for its tiny blue flowers and dots, as well as the blue trim. Marked "#4."

"GOLD CHINTZ," Leaf Celery, $300.

CHARM HOUSE, Teapot, $300.

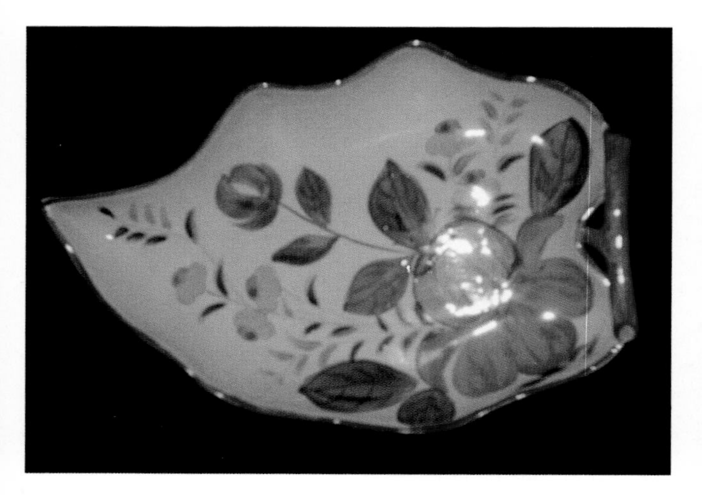

ROSE GARDEN, Mod Leaf Relish, $95.

PASQUE TULIP, 12" Loop Handle Relish, $75.

GARDEN LANE, Leaf Relish, $100. Notice that the top flower is pink. Whether it is a lavender variation or a painter mistake, it is still lovely.

JESSAMINE, Mod Leaf Relish, $100.

Above: GEORGIA, Lazy Susan, $700-$1,000. Each piece is approximately 10" long and 4" wide. Each piece is backstamped.

Left: Bottom of the GEORGIA Lazy Susan. This set appears to be in its original condition as sold in the 1950s by G.H. Specialty Co., Milwaukee, Wisconsin. Original price – less than $12.00!

EORGIA, Astor, Spoon Rest, $55.

"MIDNIGHT," China Salt and Pepper Shakers, $100.

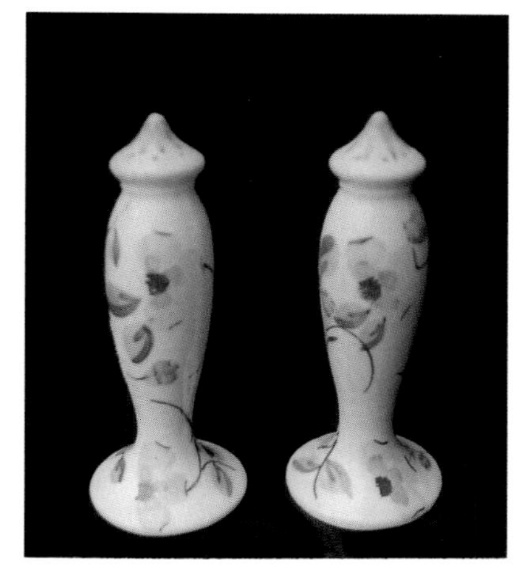

SKITTER, China Salt and Pepper Shakers, $95.

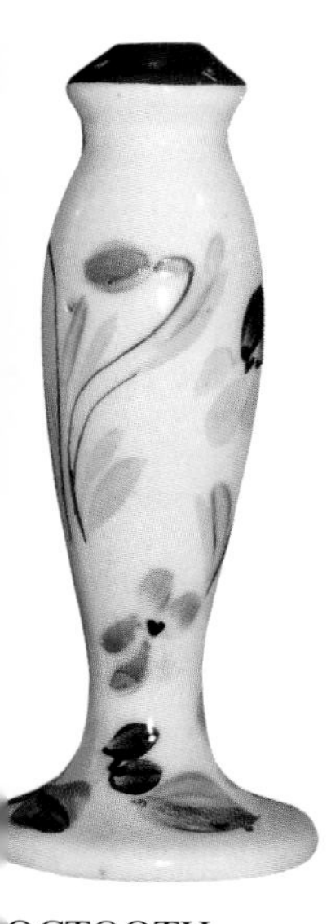

OGTOOTH IOLET, China aker. Note top of aker.

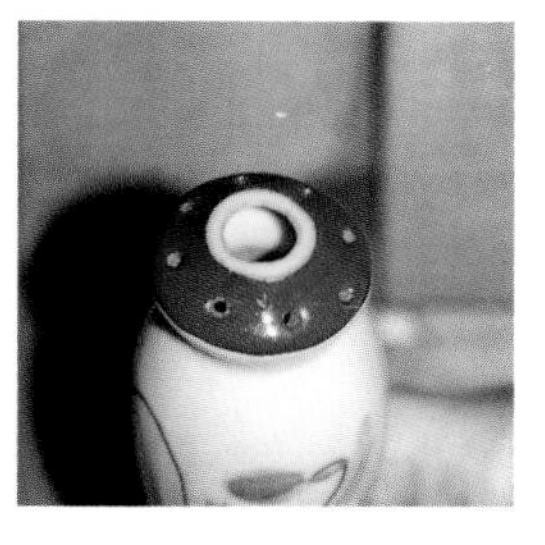

Top of DOGTOOTH VIOLET Pepper Shaker has been cut off. Shaker was sold as a "pretty, floral hat pin holder."

SWEET PEA, China Shaker, $75 pair.

"ELUSIVE," Powder Box, $250 with lid. Note variations in these two boxes. The bottom of the powder boxes. The boxes are slightly different in size because two different molds were used. Therefore, purchasing tops and bottoms separately may not guarantee a match.

ottom of the PEACOCK Candy Box. Most these boxes are stamped: "Blue Ridge, hina, Hand Painted, Underglaze, Southern otteries, Inc., Made in USA."

ROSE MARIE, Candy Box, $200. Candy boxes are softly rounded and a little less than 2" deep.

AMERICAN BEAUTY, Sherman Lily Box, $500. The lily has been found on many patterns, including DIMITY, ANNIVERSARY SONG, and FALL COLORS.

Square Boxes and Ashtrays come in many patterns.

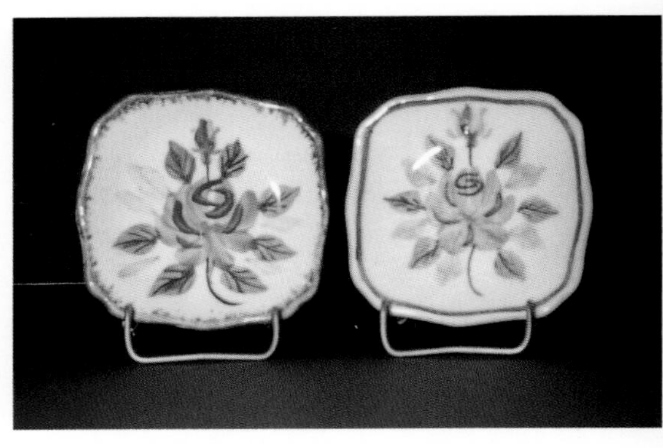

"MARIE'S ROSE" and "MICHAEL'S ROSE," Ashtrays, $25 each.

NAUGHTY, Individual Ashtray, $20. PINK DAISY, Individual Ashtray, $25.

TULIP, Ashtray, $20. Four of these individual ashtrays came inside the square boxes.

BUTTERFLY, Individual Ashtray, $25. WATERLILY, Butter Pat, $50.

"POWDER PUFFS," Square Box, $95. Matching Individual Ash Tray, $20. Note how the ashtrays often repeated only a small bud of the larger pattern.

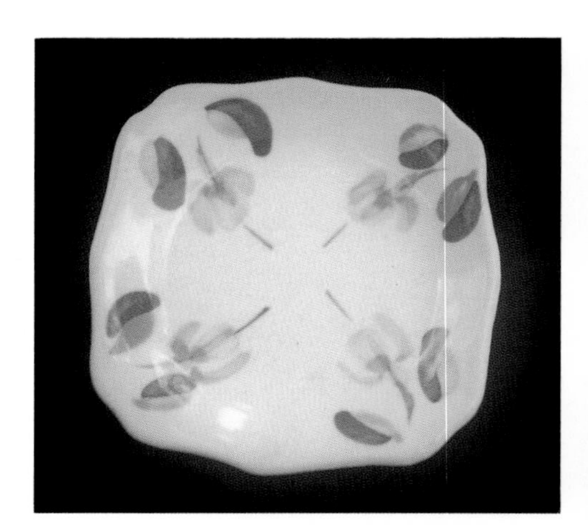

FOUR CORNER ROSE, Individual Ashtray, $20.

GAILEY, Square Box, $100, Ashtray, $25.

ottom of MALLARD DUCK box, measures 4" x .5" x 1.5". WHIG ROSE, Individual Ashtray, easures 2" x 3". Two of these ashtrays fit into the ectangular boxes.

WHIG ROSE, Individual Ashtray, $45.

A fine collection! Note MALLARD DUCK BOX on top shelf.

BURGUNDY BETSY," etsy Jug, $125.

"FLOWERS OF BLUE," Betsy Jug, $150.

"MISS LIME," Betsy Jug, $125.

andlewick Shape

EMPRESS," Platter, $25. Backstamp ncludes the name EMPRESS and 22 kt old Decorated.

"NASCO PRINCESS," Cup and Saucer, $25. Backstamp on saucer shows NASCO jobber mark and PRINCESS pattern name.

"NASCO PRINCESS," Dinner Plate, $25. These decal plates were made with a variety of different colored borders.

"PASADENA," Bread and Butter Plate, $7. This plate was made for Nasco and has the name on the backstamp. It is often found with advertising.

LITTLE GIRL, Vegetable Bowl, $25. Dinner Plate, $20.

Willene and Glenn Clark with PAULA'S LEI, named for their daughter.

TOGETHER, 9" Plate, $25.

Westwood backstamp on TOGETHER Plate.

RED BANK, Cup and Saucer, $25.

PAULA'S LEI, Covered Vegetable Dish, $75. Usually found with a red line.

WATERLILY was chosen to display this gourmet meal. (Source Unknown)

BLUEBELL BOUQUET, Set of Eight-Ounce Glasses.

BLUEBELL BOUQUET (GREEN LEAF), Cake Lifter, $30.

BOUQUET (BLUEBELL BOUQUET, GREEN LEAF), BLACK-EYED SUSAN, LIGHT-HEARTED, and ASTER (CHRYSANTHEMUM) advertised in ALLIED HOUSE BEAUTIFUL, October, 1948. It is interesting to note the labeling of pieces and prices. The 32-piece service for 6 sold for $9.95 or $10.95 west of the Rockies.

MARJORIE, Salad Bowl and Underplate, $125.

"PINK CROCUS," Bread and Butter Plate, $5.

BLUEBELL BOUQUET (YELLOW LEAF), Ovide, Coffee Pot, $200.

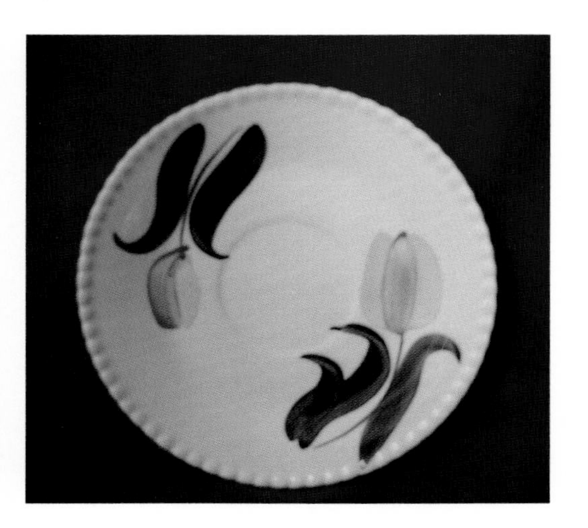

"FALL CROCUS," Saucer, $4. BROWNIE is a color variation of this pattern.

LIZZIE'S GIFT, 7" Plate, $10.

"WILLENE," 11" Cake Plate, $45.

WILDWOOD FLOWER, Salad Bowl and Underplate, $125. Note Candlewick edge around salad bowls.

ARLENE WITH LEAF, Salad Bowl, $40.

KAYLA RUTH, Dinner Plate, $25.

ARLENE, Platter, $25. This pattern was used often on granny bowls with different edges and shapes.

RECIOUS, Dinner Plate, $25.
up and Saucer, $20.

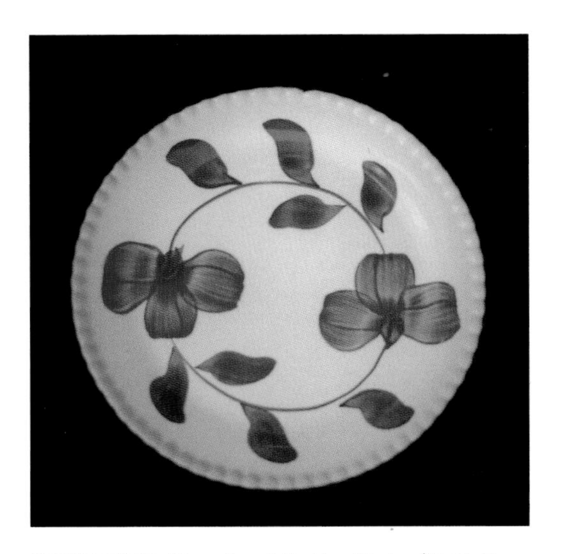

KELLY, 9" Vegetable Bowl, $25.

DUPLICATE, Bread and Butter Plate, $5. Also found on Astor and Colonial shapes.

ROSE RED, Bread and Butter Plate, $5. Full pattern has three flowers and forms a circle.

Country hutch combining stuffed rabbits, model trains, and, of course, Blue Ridge plates.

ROSEMARY, Dinner Plates, $25 each. Note the difference in painting.

ALEXANDRIA, Dinner Plate, $25.

RED CARPET, Dinner Plate, $25. Also found on Astor shape.

OBION, Dinner Plate, $20.

ɅICHELLE, Dinner Plate, $25. There is a Lace
:dge bowl also named MICHELLE.

POPPY DUET, Vegetable Bowl, $25. Shown with matching glasses.

BETTY table setting at a church luncheon. Each table is set in a different Blue Ridge pattern.

CALIFORNIA POPPY, Dinner Plate, $20.

DUFFIELD, Dinner Plate, $20.

PRETTY PETALS, Vegetable Bowl, $25.

GRANDMOTHER'S PRIDE, Cup, $8. PRIMULA is a very similar pattern.

ADORATION, Bread and Butter Plate, $5. Full pattern has three flowers.

GERBER DAISY, Plate, $15.

SUNGOLD #1, Ovide, Coffee Pot, $150.

YELLOW DREAMS, 10" Plate, $20.

ANNE ELAINE, Fruit Bowl, $8. Full pattern also has a red flower.

"COLETTE," Saucer, $5. The name "COLETTE" appears on the backstamp. This pattern on the Colonial shape with a red edge line is called PETUNIA or LAUREL WREATH.

Colette backstamp.

.ICIA, Creamer, $20. Also known as
OUNTAIN MEADOWS.

BLUE MOON, Dinner Plate, $25. This plate has the traditional Blue Ridge backstamp. However, an almost identical pattern has been found with the name BLUE BLOSSOM indicated on the back.

BRACELET, Vegetable Bowl, $30.

)ORATION, Two-Tier Server, $45.

RED TULIP, Cake Set, $95.

M POM, 20-Piece Set in the original Southern Potteries, Inc.
pping box, $300.

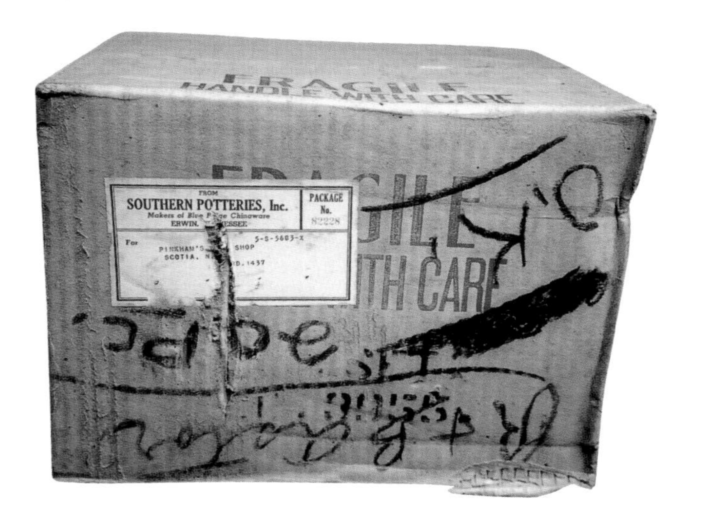

Southern Potteries, Inc., Makers of Blue Ridge Chinaware, Shipping Box, 20 pieces of #3056. Package No. 8228 was shipped to Pinkham's Gift Shop, Scotia, New York.

"GWEN," Bread and Butter Plate, $5. Full pattern also has a yellow flower.

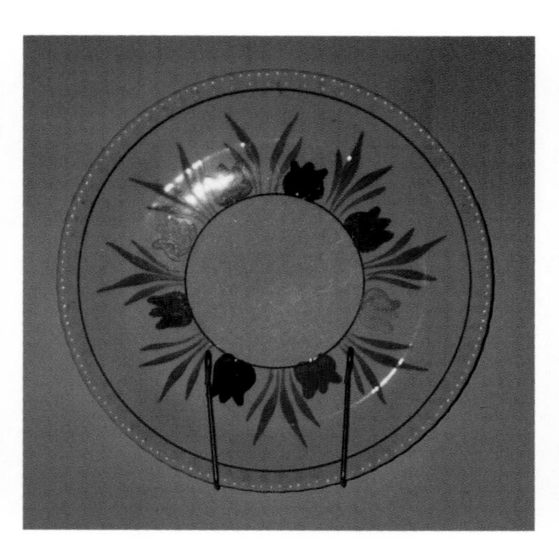

TULIP GARDEN, Dinner Plate, $25.

PEGGIE'S POSIES, Plate, $20.

"TULIP SCROLLS," Candlewick, 9" Bowl, $30. This pattern with the Golden Rule is called MOTTO WITH SNOWFLAKE.

RED TULIP, Bread and Butter Plates, $10 each. One plate has the Franklin Kent backstamp; the other Blue Ridge. One measures 5.25" and the other 5".

"B.J.," 6" Plate, $5. Similar to CHATHAM.

Glasses to go with ALLEGRO (VIBRANT).

TULIP TREAT, 6" Square Plate, $25. PV backstamp on this plate. An almost identical plate has been found with a French backstamp.

"PEGGIE'S POSIES VARIANT," Fruit Bowl, $8.

(Left to right) "ST. PATRICK'S FLOWER," 6" Plate, $4. "PASTEL TULIP CIRCLE," 6" Plate, $5.

TWIN TULIPS, Dinner Plate, $20.

"ALEENA VARIANT," Dinner Plate, $15. Note the reversal of the flowers. UCAGO backstamp.

FULL BLOOM, Cup, $10. Full pattern includes half pink flower.

vable GUMDROP TREE tucked uggly into a nightstand.

MOCCASIN, 9" Plate, $20.

CUMBERLAND and MOUNTAIN IVY, with matching glasses, as advertised in the Spring & Summer Sears catalog, 1948.

HEIRLOOM, Covered Vegetable, $75. Has also been called CONFETTI.

LANGUAGE OF FLOWERS, Authors' complete set of 8" plates. Due to its scarcity, the LANGUAGE OF FLOWERS set has increased in value to $150 per plate or $1,000 for the complete set.

LUELLA, Dinner Plate, $25.

"SQUIGGLE," Dinner Plate, $20. WIGGLE (Colonial) is very similar pattern.

MAPLE LEAF RAG, Salad Bowl, $45.

OAKDALE, Bread and Butter Plate, $8. Also found on Colonial shape.

GREEN LANTERNS, Creamer, $20.

Piecrust Shape

Blue Ridge heaven for collectors!

SPRING MORNING, Bread and Butter Plate, $5.

EMLYN, Platter, $30. Dinner Plate, $20.

SPRING BLOSSOM (CLINCH MOUNTAIN DOGWOOD), Teapot, $125 with lid.

VELMA, Dinner Plate, $40.

APPALACHIAN GARDEN, Dinner Plate, $25.

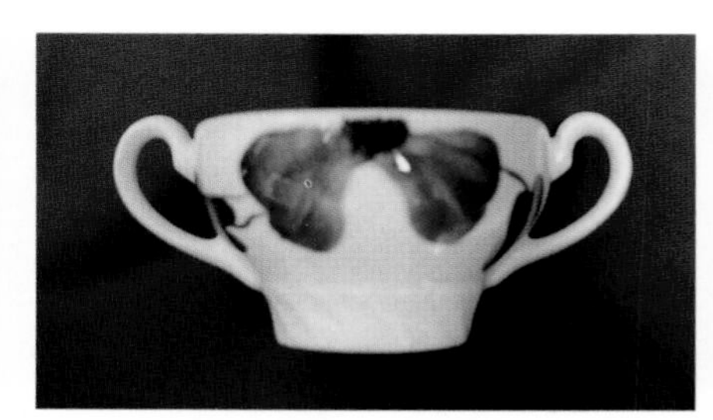

CLAREMONT, Open Sugar, $20.

The other side of CLAREMONT Open Sugar.

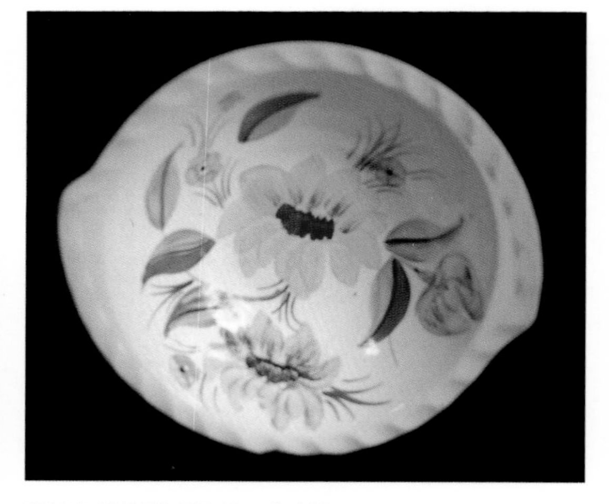

CALLAWAY, Tab Bowl, $25.

TRUMPET VINE, Luncheon Plate, $20.

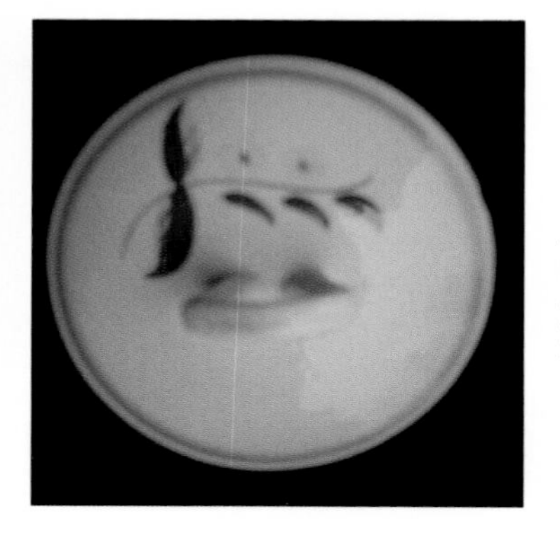

JESSAMINE, Teapot Lid, $20. Lids and teapots purchased separately may not be a match.

ED RING, Sugar and Lid, $25.

JESSAMINE, Luncheon Plate, $25.

This beautiful pitcher could be used with MAGNOLIA, SPRING BLOSSOM, SOUTHERN DOGWOOD, and several other patterns.

MAZURKA, Dinner Plate, $25.

MAGNOLIA, Teapot, $200.

Sheila Ferguson is enchanted with MAGNOLIA.

"DAYLILY," Teapot, $150.

TEMPO, Casserole with Lid, $55.

Glasses in three sizes coordinate with SOUTHERN CAMELLIA and other patterns.

SOUTHERN CAMELLIA is featured in this advertisement in the October 1984 issue of HOUSE BEAUTIFUL. Southern presents "the Beautiful New Pie Crust Shape."

LOTUS, Cup, $15.

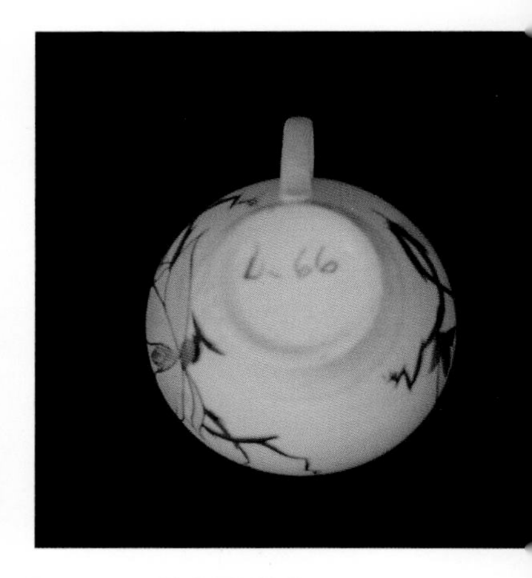

Bottom of LOTUS Cup, marked L-66.

BACKYARD BLEEDING HEART, Sample Cup, $50.

BACKYARD BLEEDING HEART, Dinner Plate, $25.

Inside of Cup showing Sample #4030 Approved.

BOUTONNIERE, Fruit Bowl, $5.

SECRET GARDEN, Dinner Plate, $25. STUCK UP is a similar pattern.

RIDGE IVY, CRAB APPLE, GREEN BRIAR, COUNTRY CHARM (RED BARN), and SUN BOUQUET in a Montgomery Ward catalog, Fall & Winter, 1952-53. Note the glasses in three sizes made to match these patterns.

SECRET GARDEN, Cake Lifter, $25. Lifter could be used with other patterns.

Glass pitcher with GREEN BRIAR-type flowers could be used with many patterns. Matching glassware made by several companies was sold through Sears and Wards catalogs.

KIBLER, Ovide, Coffee Pot, $125.

GREEN BRIAR, Ovide, Coffee Pot, $150. Matching glasses in three sizes were probably made by Federal.

Skyline, Trailway, and Palisades Shapes

"GLENN," bowl, $15.

SWAMP MALLOW, Cup and Saucer, $25.

"MARY ANN," Dinner Plate, $25.

"MOSS ROSE VARIANT," Dinner Plate, $15.

"LANA," Dinner Plate, $25.

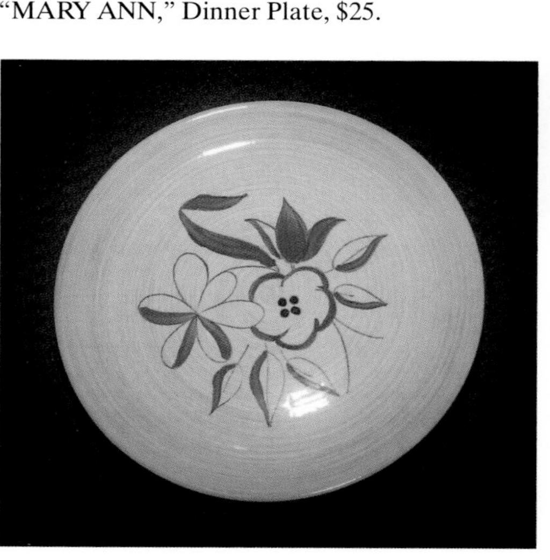

Above: TWIG, Barrel Salt and Pepper Shakers, $40. Also matches ARTFUL.

Left: BARDSTOWN, Dinner Plate, $20.

Right: GLORIA JEAN, Vegetable Bowl, $20.

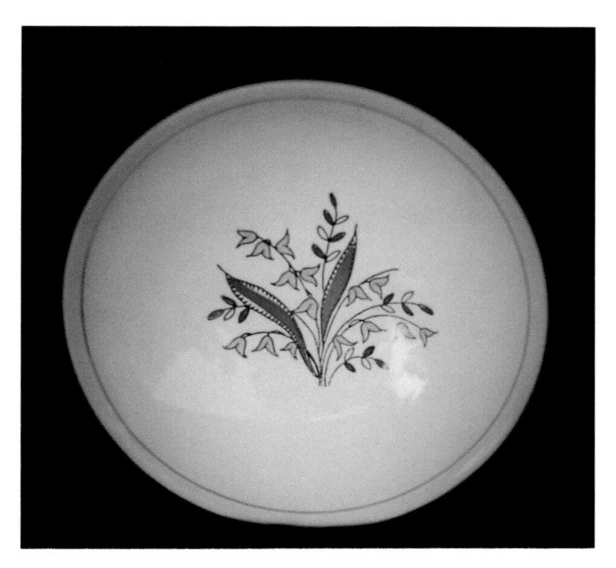

TWIG, Vegetable Bowl, $25.

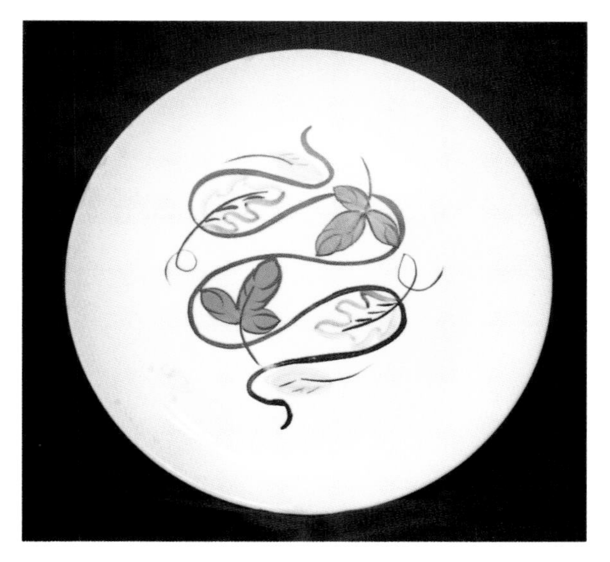

"AUTUMN BREEZE," Dinner Plate, $15.

ONAIRE, Dinner Plate, $15.

SOUTHERN STARBURST, Dinner Plate, $15.

BRIAN, Dinner Plate, $15. #4633-F.

SCARLET LEAVES, Clock, $35. Full pattern has three leaves.

ALOHA, Vegetable Bowl, $20.

"VENUS," 9" Plate, $15.

ATHENS, Dinner Plate, $20.

Back of ATHENS plate, marking it Tem. #593.

"GUSTY," 7" Plate, $5. Pattern has pink streaks across the background.

GLORY, Dinner Plate, $20.

RAZZLE DAZZLE, Big Cup, $25. This pattern in aqua is called UGLEE.

PINKIE LEE, 6" Tile, $40. Tiles are seldom marked. An almost identical pattern with paler flowers is PINK CHIFFON.

TINY, Rope Handle, Cup, $5. Saucers have a red flower; full pattern has both.

TEA ROSE, Rope Handle, Creamer, $20. Rope Handle shapes were popular grocery store premiums.

ARK, 4" Big Cup, $45.

DAMASCUS, Bread and Butter Plate, $5. Dinner plate also includes a blue flower.

"KIND OCTOBER," Egg Cup, $30. Matching blue-gray line appears inside cup. ORCHID, Cup, $5.

ORCHID, Square Plate, $25.

IARLESTON, Bread and Butter Plate, $5. Two flowers on the full pattern ve six petals. TWIN TULIPS, Candlewick, Bread and Butter Plate, $5.

BROWN DAISY, Fruit Bowl, $5. This is a color reversal of DESERT FLOWER.

CECELIA, Divided Vegetable Bowl, $25.

"NORTHERN DOGWOOD," Dinner Plate, $10. #4174 stamped on back. #4175 is SOUTHERN DOGWOOD.

MARIETTA, 10" Plate, $15.

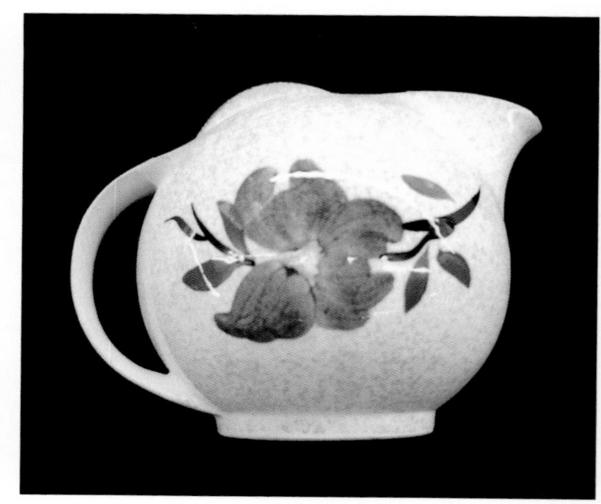

CAROLINE, Teapot, $75.

ORANGE QUEEN, Skyline Teapot, $75.

"SPIDER VIOLETS," Big Cup, $60. SPIDERWEB, Big Saucer, $30. This is the pattern used for the "FATHER" cup.

VALDOSA, Rope Handle Teapot, $165. Rope Handle Teapots, which usually complement Skyline patterns, are rare.

PRIM, 9" Coffee Carafe, $95. This popular fifties piece was originally sold with a metal stand with a candle or Sterno insert.

Stetson pitcher and creamer often confused with Blue Ridge.

"CURLY," 9" Plate, $25.

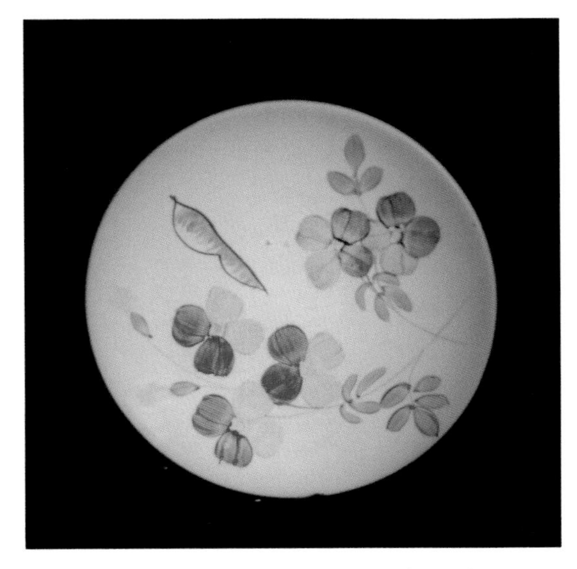

WILD GERANIUM, Dessert Plate, $7. Full pattern had the additional large round green leaf.

CHEROKEE ROSE, Trailway, Dinner Plate, #20. Also found on Skyline shape with no border. Look for Rope Handle accessory pieces.

"VIOLET BLUSH," Dinner Plate, $25.

PIANISSIMO, Cup and Saucer, $20.

CHEERIO (YELLOW), Skyline, Teapot, $75 with matching lid.

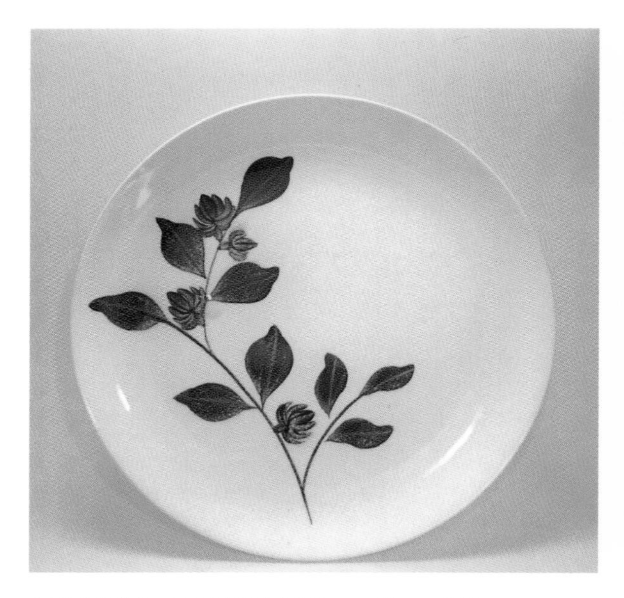

CAROLINA ALLSPICE, Dinner Plate, $20.

"CONRO," Dinner Plate, $20. Westfall backstamp.

DOE EYES, Square Plate, $10.

SUNNY SPRAY is described as "cheerful as sunlight" in this advertisement.

"WINNIE'S MOUNTAIN DAISY," Dinner Plate, $30.

HEERIO (YELLOW), Covered Toast, $100. The vered toast lid fits the square plate, but is more cure on the 8" round plate with 6" indentation.

LEAF SPRAY, Salad Plate, $8.

NIGHT FLOWER and GOLDEN SAND (GOLD PATRICIA) are "fascinating new Blue Ridge patterns" as this advertisement proclaims.

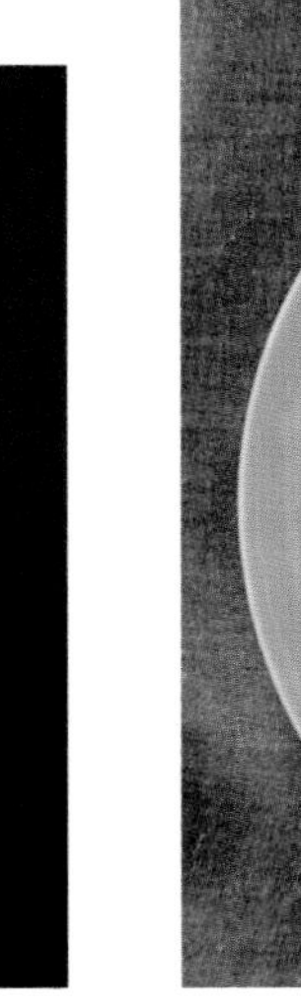

PRAY AROUND," Plate, $10.

"KATHLEEN," Dinner Plate, $20. Identified on back as Tem. #101-D.

Back of "KATHLEEN" plate, marking it Tem. #101-D.

Above: MIDAS TOUCH, Custard Cup, $25. Also called SHERBET.

Left: DAISY GOLD, Oval Bowl, $20. This is a Sears 1950s pattern.

MEADOWLEA, 10" Plate, $20.

"TENNESSEE PINE," Dinner Plate, $120.

Back of "TENNESSEE PINE" Plate, indicating TEM #.

PARTRIDGE BERRY, Skyline, Fruit Bowl, $5.

Stetson Dogwood and Stetson Daisy patterns are often mistaken for Blue Ridge.

A collection of custard cups. (Left to right, top to bottom) CADENZA, "TYLER," SUN BOUQUET, CAMELOT, and FLAPPER. A totally different pattern on Piecrust is also called CAMELOT. These cups were listed in the 1950s for $1 each.

SUNBURST, Dinner Plate, $20. This pattern was named by Winnie Keillor. A similar pattern has subsequently been named PINCUSHION. They are found with different "eye" treatments.

BRIARWOOD, Dinner Plate, $15.

COSMOS, 12" Platter, $20.

FLAPPER, Dinner Plate, $15. This pattern is often confused with SOUTHERN DOGWOOD.

SOUTHERN DOGWOOD, SUNNY SPRAY, and DESERT FLOWER were presented as "a beautiful bouquet of flower patterns."

CHIFFON (PINK), Trailway, Dinner Plate, $15. This pattern was also made on Palisades.

INGE, Small Mixing Bowl, $40. This pattern on Piecrust is called CANDACE.

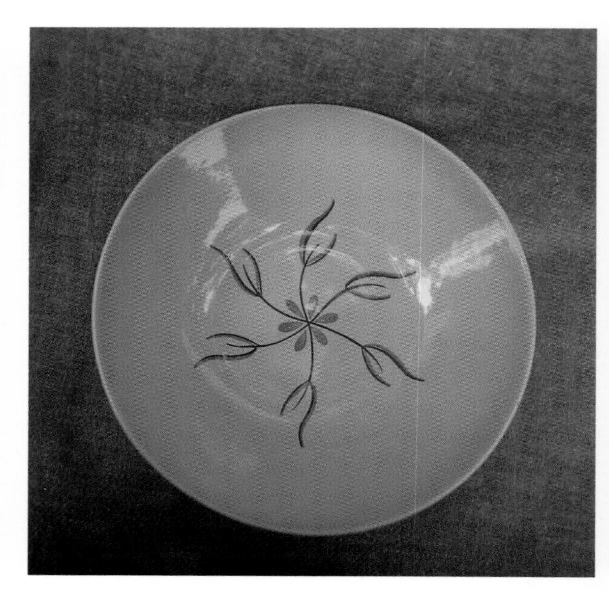

ANDANTE, Vegetable Bowl, $15.

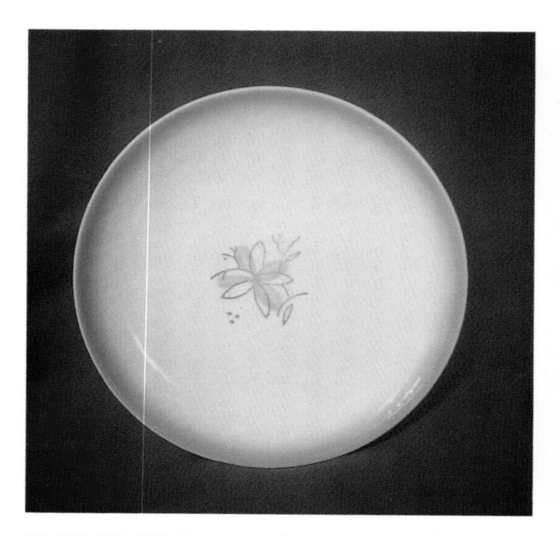

BLUE TREE, Bread and Butter Plate, $5.

"JENNIFER," Dinner Plate, $15.

"ELBERTON," Dinner Plate, $10.

BERRY PATCH, 10" Plate, $20.

KISMET, 10" Plate, $15. A totally different pattern on Colonial shape also has the name KISMET.

"BILL," Platter, $15.

MISSISSIPPI, Platter, $20

"HENRY," Fruit Bowl, $5

MODERN LEAF, Ovide, Coffee Pot, $100.

"RINGED IVY," Saucer, $4.

"PYRACANTHA VARIANT," Dinner Plate, $15.

WHIRL, Bread and Butter Plate, $5.

GRAPE IVY, Plate, $15.

DUNGANNON, Dinner Plate, $15. The perky little yellow flowers played hide and seek with our cameras.

SHEPHERD'S PURSE, Skyline, Teapot, $70.

SOLIDAGO, Teapot, $65.

PLANTATION IVY, advertised by Southern in the July 1951 issue of AMERICAN HOME as "When food tastes better."

"HUGH KIBLER," Skyline, Vegetable Bowl, $25.

WILLA, Dinner Plate, $15.

STANHOME IVY, Skyline, Cup, $5.

Southern Pottery box used for shipping a twenty piece set ($12.95). Date on box is October 30, 1952.

The set of STANHOME IVY in this box was not made by Southern Potteries. Note the shape (Stetson) of the cup handle. The box is also dated 1959, two years subsequent to the closure of Southern.

"SPRIG VARIANT," Celery or Gravy Stand, $15. This pattern has also been found on the Trailway shape with some reversal of colors.

COLUMBINE, Dinner Plate, $25.

HUGH KIBBLER," Palisades, itcher, $75.

LEAVES OF FALL, Tab Bowl, $15. Also on Trailway shape.

Sandy Ligerfelt with her ROBIN Platter.

GRAY DAY, Palisades, 15" Platter, $25.

GARDEN PINKS, Platter, $20.

"PINK CHAMPAGNE," Tab Bowl, $10. Backstamp is Westfall.

CHIFFON, Palisades, Pitcher, $85.

BERRY PATCH, Palisades, Plate, $20. Another Skyline pattern which has been repeated on Palisades.

STARFLOWER, Vegetable Bowl, $20. "GOLDEN STACCATO" is a similar pattern with turquoise leaves.

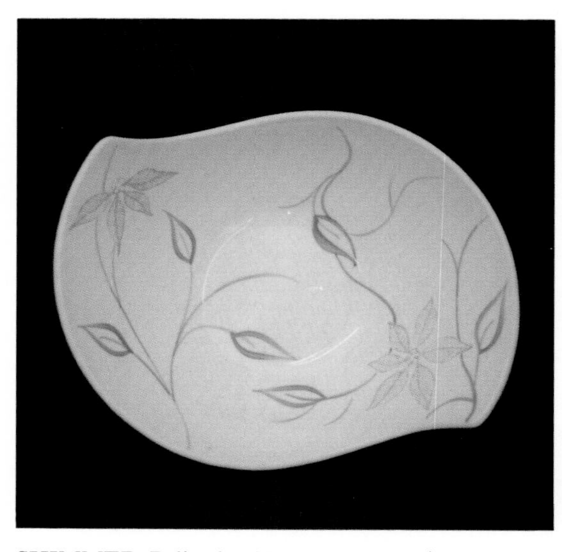

SHIMMER, Palisades, Vegetable Bowl, $25.

STACCATO, Cup and Snack Plate, $30. This pattern with five straight petals is called SPOKES.

RUE DE LA PAIX, Skyline Lid for Rope Handle Sugar, $5.

BLUEFIELD, Vegetable Bowl, $20. The little blue flowers give this pattern its name. Another pattern, also named BLUEFIELD, has a large blue nocturne-type flower.

MAYFLOWER with blue, Palisades, Platter, $25.

FRENCH KNOT, Palisades, Cup and Saucer, $10.

"TYLER," Palisades, Pitcher, $85.

Woodcrest Shape

BALI HI (BAMBOO), CRAB APPLE, CALICO (COUNTRY FRUIT), and RIDGE IVY are featured in a Montgomery Ward Fall & Winter catalog, 1954-55.

MIMIC, Dinner Plate, $25.

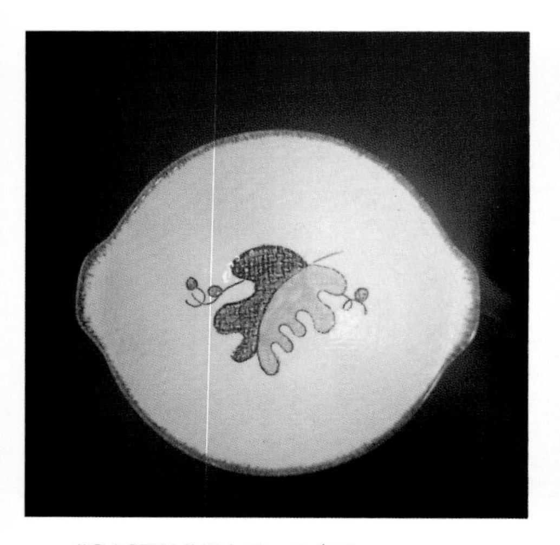

"CASEY," Tab Bowl, $15.

SWEET ROCKET, Bread and Butter Plate, $10.

CARETTA CATTAIL, 9" Plate, $20.

MOUNTAIN GLORY, Dinner Plate, $25. This pattern does not have the yellow line to distinguish it from YELLOW NOCTURNE.

QUILTED IVY, Teapot, $125 with lid.

EXOTIC, Oval Bowl, $25.

Jami Willoughby, great, great granddaughter of Novella Beals, with MING TREE Creamer.

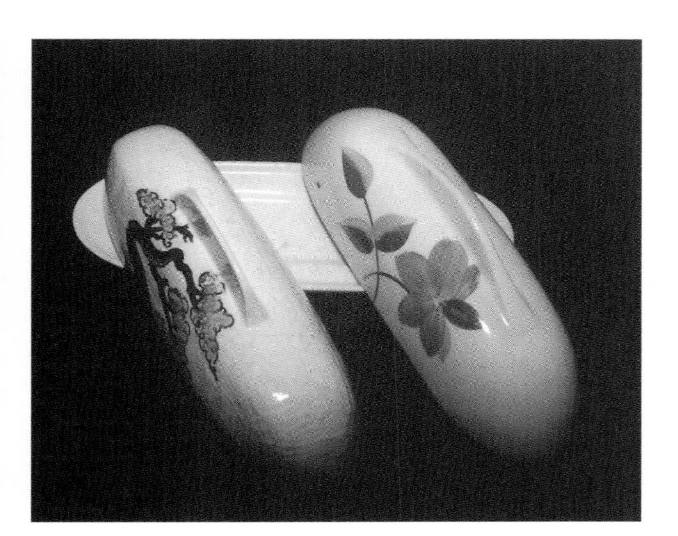

MING TREE, Woodcrest, Butter Dish Lid, SUSAN, Skyline, Butter Dish Lid. The finial on top of the Woodcrest lid is less curved, perhaps in an effort to make it easier to hold.

ARABELLA, Teapot, $125. This pattern is usually found on Colonial.

Chapter Four
Fruit and Vegetable Patterns

Flower and Fruit Patterns

Not enough patterns combining flowers and fruit were produced to satisfy the demands of today's collectors, many of whom are acquiring sets for everyday use.

MARY, Candlewick, 11" Cake Plate, $75.

CHERRY DROPS, Cup and Saucer, $20.

MARY, Ovide, Coffee Pot, $250.

Marion Henry enjoys using her set of DELLA ROBIA.

DELLA ROBBIA, Celery Dish, $25.

ERIN, Clinchfield, Dinner Plate, $35. Patterns that combine fruits and flowers are very popular with collectors.

"LEMON YELLOW," Saucer, $5. An additional fruit and flower pattern to search for.

CHERRY BLOSSOM, Bread and Butter Plate, $10. UCAGO backstamp. (*See Appendix for an explanation of UCAGO.)

Fruit Patterns

American kitchens are once more decorated with wallpaper, tiles, and linens in fruit motifs. Fruit motifs add charm to American homes and increase the demand for Blue Ridge fruit patterns.

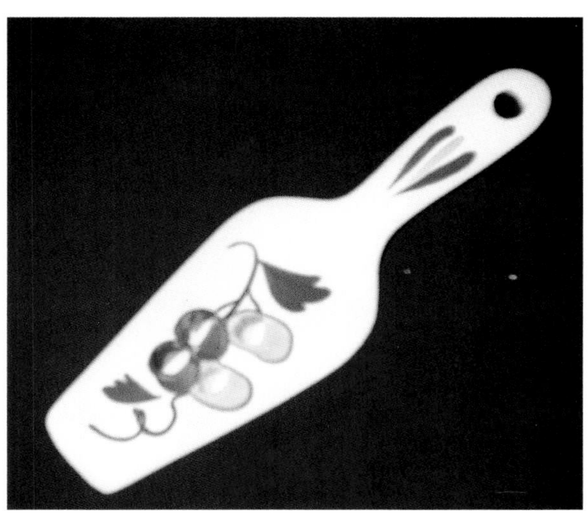

FREEDOM RING, Cake Lifter, $30.

Display case with several fruit patterns and GOLD LADY Flower Frog on the top shelf.

"RINGED FRUIT," Tab Bowl, $20.

DIXIE HARVEST, Piecrust, Sugar, $25 with lid.

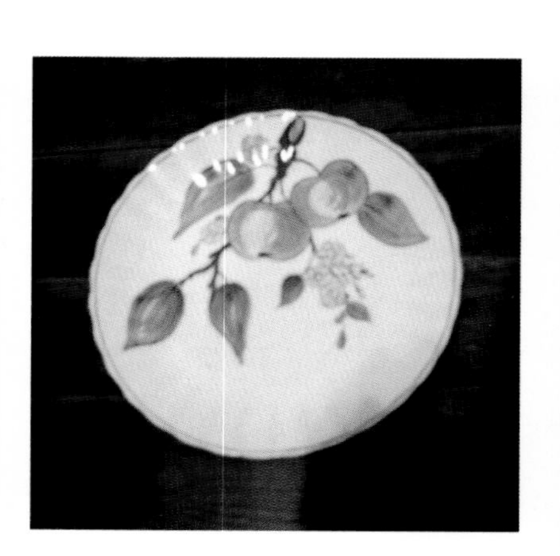

"GEORGIA PEACHES," Luncheon Plate, $25.

LAVENDER FRUIT, Colonial, Fruit Bowl, $8.

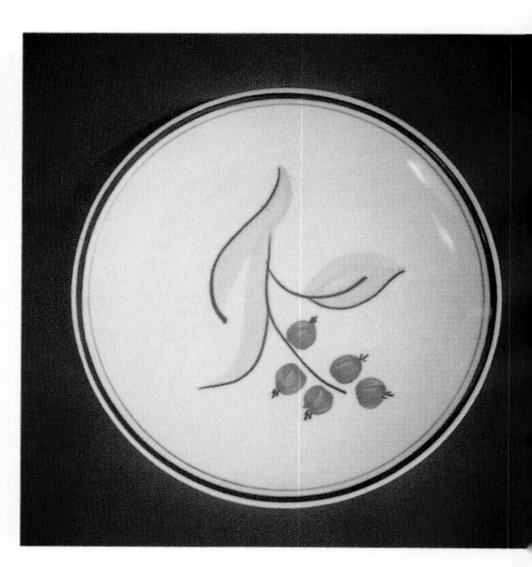

PIPPA, Bread and Butter Plate, $5. This pattern can be seen in the June 1951 issue of CHINA, GLASS AND DECORATIVE ACCESSORIES magazine.

GYPSY FRUIT, Colonial, Cake Plate, $35.
CALYPSO is very similar with a single line edge.

ABUNDANCE, Platter, $45.

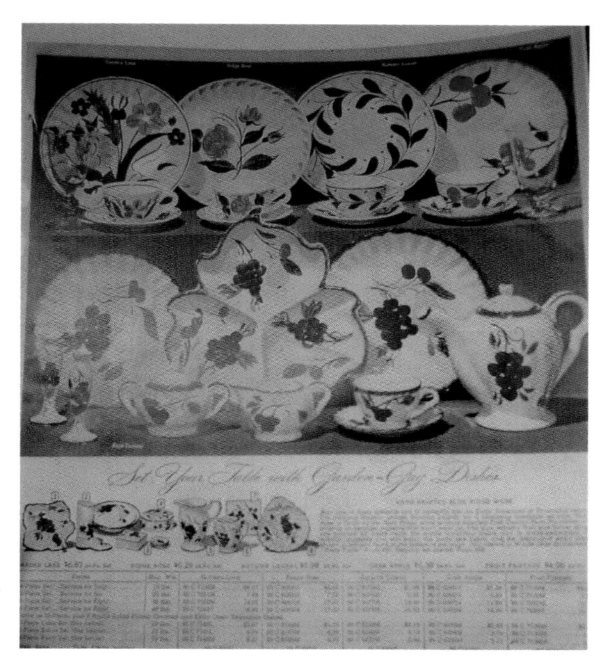

GARDEN LANE, RIDGE ROSE, AUTUMN LAUREL, CRAB APPLE, AND FRUIT FANTASY were sold by Montgomery Ward for several years. Displayed in the FRUIT FANTASY pattern are all the pieces available. It is interesting to note that the ad states that no gravy stand is available for any of the patterns. It is suggested that a bread and butter plate be used.

FOLK ART FRUIT, Square Plate, $40.

Above: FRUIT FANTASY, Cake Plate and Lifter.

Right: Back of FRUIT FANTASY Cake Plate. Inscribed: Hazel McIntosh, Nov 1, 1944, #100.

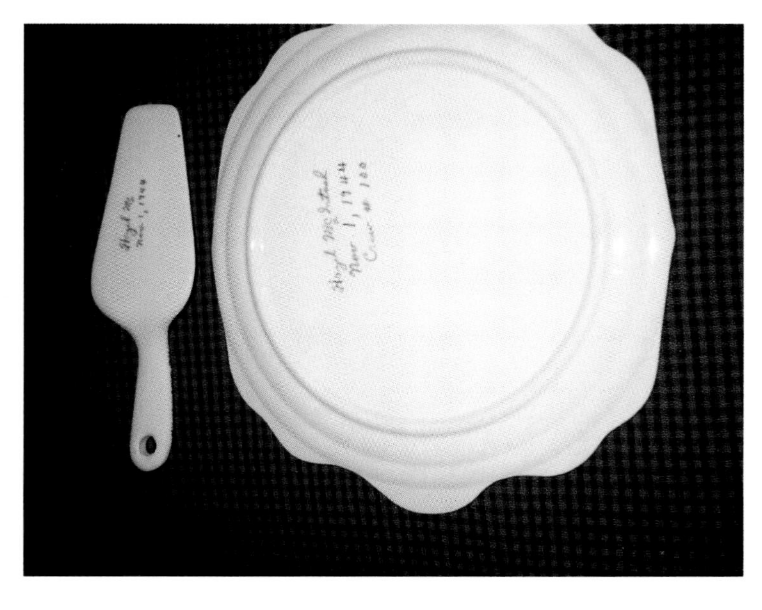

"FRESH FRUIT," Tile, $45.

CHABLIS, Egg Plate, $75. A late 1940s pattern.

SUTHERLAND, Colonial, Dinner Plate, $25.

BOUNTIFUL, Colonial, Oversized Open Sugar, $25.

(Left to right) FONDEVILLE FRUIT, FONDEVILLE PUNCH, and JASON Salad Plates, $25 each. These are marked Fondeville and may be part of another set made for Fondeville.

DANA, Candlewick, Vegetable Bowl, $25.

FRUIT MEDLEY, Candlewick, Dinner Plate, $25.

RCHARD GLORY, Ovide, Coffee Pot, $200.

"HOPE'S FRUIT," Clinchfield, Plate, $25.

BUCKETFUL, Colonial, Luncheon Plate, $20.

HAWAIIAN FRUIT, Astor, Sugar, $30. This pattern is also found on Candlewick and Trailway shapes.

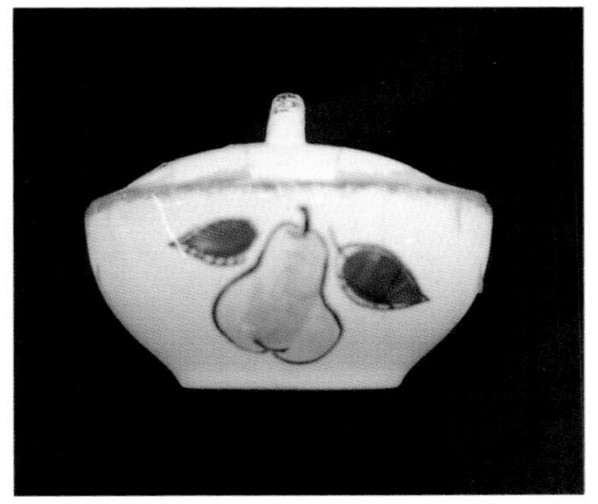

GREEN PEAR, Covered Sugar, $30. This pattern was made on Trailway and Palisades shapes.

FRUIT CRUNCH, Cake Plate, $30.

The Hudson Pulp & Paper Co., N.Y. promotional offered a four-piece starter set of "PRECIOUS PEAR" for $1.25 and one Hudson dinnerware coupon. The package advised that since the dinnerware had to be hand painted, customers needed to allow up to four weeks for delivery.

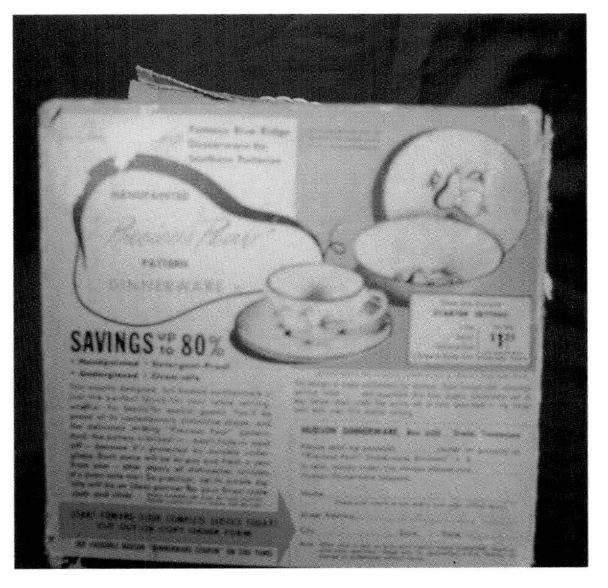

"PRECIOUS PEAR" (currently known as SHADOW FRUIT) was described as a "smart combination of black, grey and pink." Although made of earthenware, it was an ideal partner for fine dinnerware. It was oven safe and remained "fresh even after plenty of tumbles."

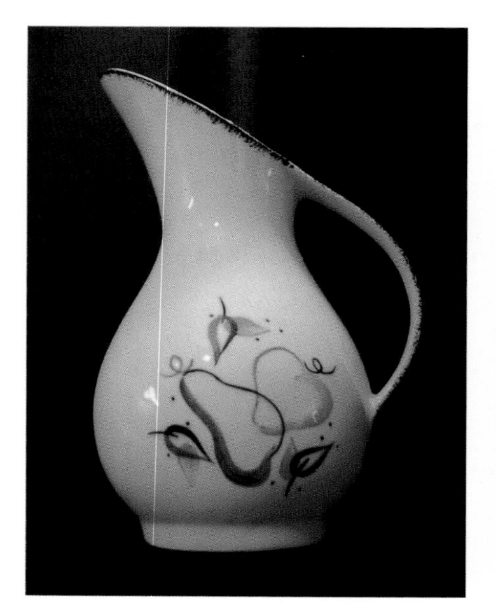

"PRECIOUS PEAR," Palisades, Pitcher, $85.

"SIMPLE FRUIT," Bread and Butter Plate, $8.

GREENCASTLE, Skyline, Fruit Bowl, $5.

"LEAFY FRUIT RING," Ball Teapot, $125. Dinner plate, $25.

Photograph of the pottery, circa 1917, surrounded by MOUNTAIN ASTOR Plate and NEEDLEPOINT FRUIT Plate.

APPLE MIX, Skyline, Divided Vegetable Bowl, $40.

APPLE AND PEAR, Woodcrest, 7" Plate, $15.

QUILTED FRUIT. The owner of this set was a Southern employee. She called this "GRASS SACK."

"QUILTED APPLE VARIANT," Woodcrest, Sugar, $40. Barrel Shakers, $35. Leaves vary on QUILTED FRUIT patterns.

GINGHAM APPLE, Skyline, Teapot, $100 with matching lid.

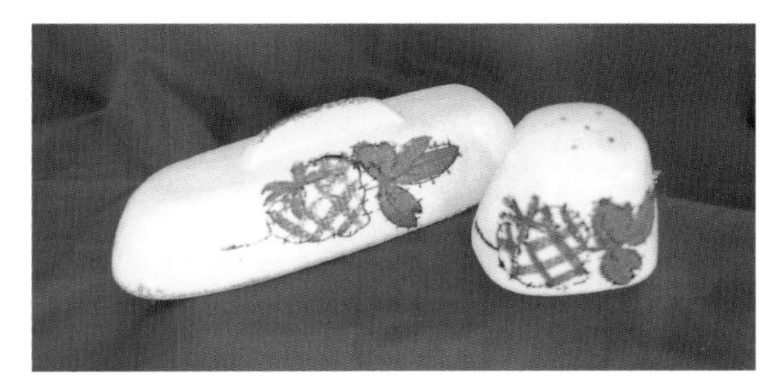

"QUILTED STRAWBERRY," Woodcrest, Butter Dish, $75 with lid. Salt and Pepper Shakers, $65 set.

GINGHAM APPLE, Skyline, Ovide, Coffee Pot, $200.

TARTAN APPLE, Woodcrest, Oval, Vegetable Bowl, $25.

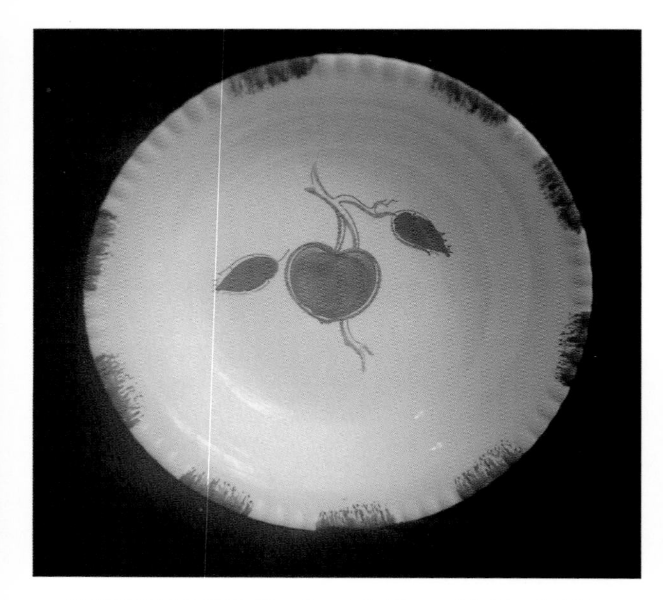

HARDY APPLE, Candlewick, Vegetable Bowl, $30.

"ODD COUPLE," Dinner Plate, $30.

"UPSIDE DOWN APPLE," Woodcrest, Salt Shaker, $40.

TARTAN FRUIT, Luncheon Plate, $25.

"UPSIDE DOWN APPLE," Woodcrest, Teapot, $200. Square Round Plate, $40. Note that the branches are not filled in on the plate, and, of course, the apple on the teapot is upside down.

BEADED APPLE, Colonial, Teapot, $125.

CRISS CROSS, Piecrust, Plate, $25. Saucer, $5. It is unusual for a complete pattern to be painted on smaller pieces, such as this saucer.

WHIPSTITCH, Piecrust, Vegetable Bowl, $25.

Bottom of WHIPSTITCH Bowl with original tag, denoting pattern decoration #4733.

APPLE TRIO, Cake Set, $125.

(Left to right) JUNE APPLE, Woodcrest, Cup, $7. ARLINGTON APPLE, Rope Handle, Creamer, $20.

It took many oatmeal breakfasts to collect this much QUAKER APPLE.

"TWIRLY APPLE," Skyline, Dinner Plate, $25.

APPLE JACK (WHITE), Three-Tier Snack Server, $75.

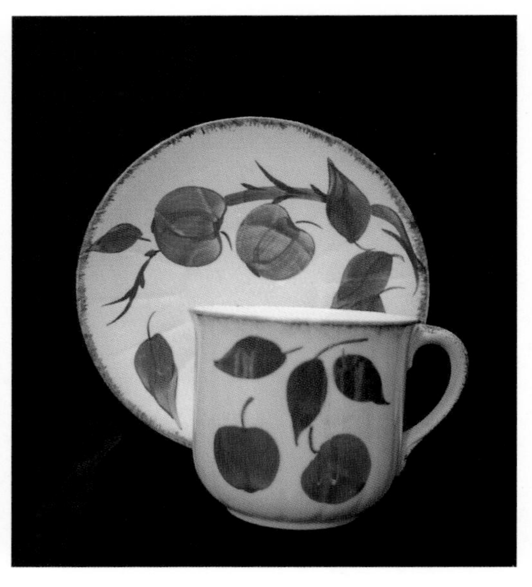

"BAY APPLES," Big Cup and Saucer, $70.

"APPLESAUCE," Colonial, Tab Bowl, $25.

RED APPLE, Candlewick, Baking Dish, $125.

WHIP TAC, Vegetable Bowl, $25.

Egg cups with eggs!

Stamp used for unknown pattern #14851-1.

APPLE STRUDEL, Salad Bowl and Underplate, $125 set.

GRANNY SMITH APPLE, Skyline, Gravy, $35.

GRANNY SMITH was "a brilliant addition to the famous Skyline family." Also shown: BETHANY BERRY, CHEERIO, SCATTER PLAID, MAYFLOWER, and SOUTHERN DOGWOOD.

"AUSTIN APPLE," Skyline, Dinner Plate, $25.

APPLE BUTTER, 9" Plate, $20.

SWEETIE PIE, Dinner Plate, $22.

WESTFALL CHINA CO. backstamp on APPLE BUTTER Plate. Westfall China Co. was a jobber.

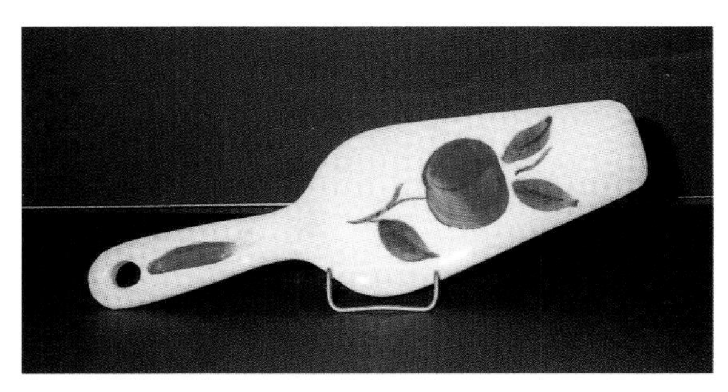

"APPLE," Cake Lifter, $35. Lifter complements several apple patterns.

CRAB APPLE is prominently displayed in this Kansas City Montgomery Ward catalog page. GARDEN LANE and RIDGE DAISY are also shown. It would be interesting to note if these patterns are more readily available today in the Kansas City area.

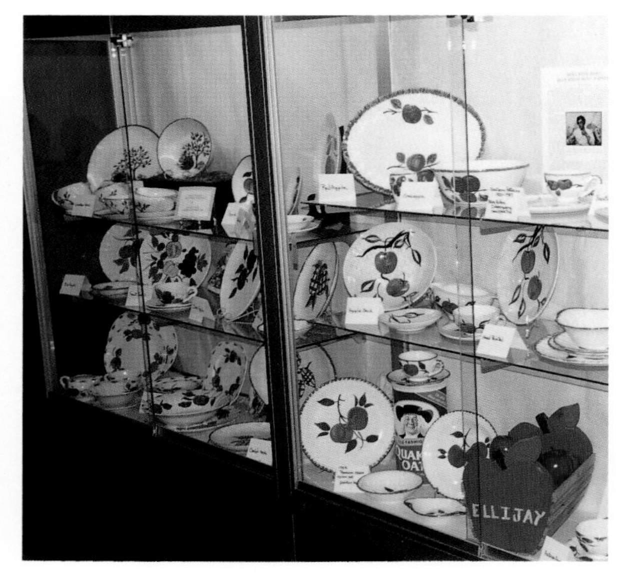

Library display during an apple festival in Ellijay, Georgia.

APPLE CRUNCH, Sample Cup, $50.

Inside of APPLE CRUNCH cup showing Sample #4309 Approved.

Unusual CRAB APPLE Wall Lamp.

JOE'S APPLES, Skyline, Dinner Plate, $30.

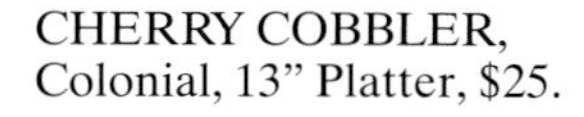

CHERRY COBBLER, Colonial, 13" Platter, $25.

BEADED CHERRY, 9" Plate, $25.

SINGLETON, Colonial, Vegetable Bowl, $20.

HILLARY, Candlewick, Vegetable Bowl, $35. Sometimes found on Lace Edge shape.

"CHERRY CROSS," 8" Bowl, $25.

MOUNTAIN CHERRY, Candlewick, Cup, $7.

DOUBLE CHERRY, Dinner Plate, $25.

CHERRY TREE GLEN, Piecrust, Sugar, $25 with lid. Also called CHERRY CLUSTERS.

WILD STRAWBERRY, Jane Pitcher, $125.

WILD STRAWBERRY, Covered Vegetable in Aluminum Stand, $95.

"SIMPLY WILD CHERRY" (#1), Dinner Plate, $20.

WILD STRAWBERRY, Clinchfield, 17" Platter, $95.
The Clinchfield Platter here is a rare piece.

WILD STRAWBERRY place setting in Better Homes & Gardens, February 1948. Pattern can also be seen in Country Folk Art, June 1992.

"EARTHY WILD CHERRY" (#3), Piecrust, Teapot, $100.

This elaborate collection of WILD STRAWBERRY includes a much sought after Alice pitcher.

WILD STRAWBERRY, Ovide, Coffee Pot, $200.

ıtique hutch with an assortment of
ERRYVILLE pieces. Often confused
th WILD STRAWBERRY even
ough the edges are different colors.

FRAGERIA, Skyline, Divided Dish, $45.

SWEET STRAWBERRIES, Skyline, Creamer, $20.

WEETHEART, 9" Dinner Plate, $20.

STRAWBERRY SUNDAE, Colonial, Teapot, $150.

BERRYVILLE, Ovide, Coffee Pot, $200.

STRAWBERRY SUNDAE, Skyline, Dinner Plate, $30.

STRAWBERRY SUNDAE, Colonial, Saucer, $5. "CLEARLY STRAWBERRIES," Clinchfield, 6" Plate, $10.

STRAWBERRY PLANT, Skyline, Cup and Saucer, $25.

STRAWBERRY PLANT, Dinner Plate, $25.

STRAWBERRY DUET, Skyline, 9" Plate, $25.

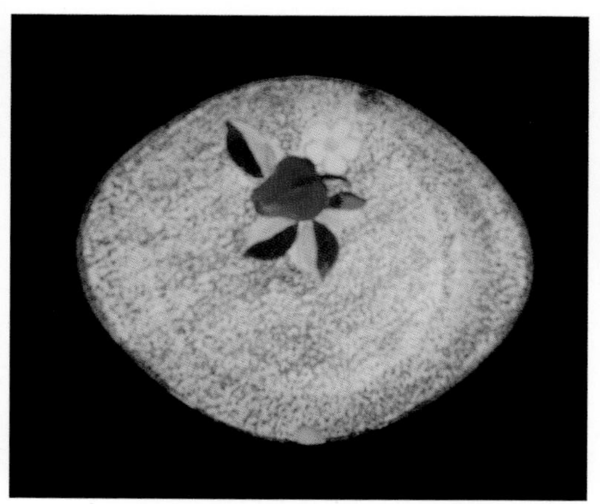

STRAWBERRY DUET, Tab Bowl, $25.

STRAWBERRY GARDEN, Astor, Bread and Butter Plate, $8.

"HARVEY'S FRUIT," Candlewick, Dinner Plate. LACE & LINES pattern in background. This unusual plate is signed E. Woodland.

JUBILEE FRUIT SET, Colonial, Teapot, $125.

JUBILEE FRUIT SET (STRAWBERRY PATCH), Cup and Saucer, $45.

JUBILEE FRUIT SET, Alice Pitcher, $150.

'ETITE FRUITS, Snack Plates, $25 each. (Left to right) "PETITE \PPLES," THOMPSON GRAPES, PURPLE PLUMS, and "PETITE 'EARS." PORTIA'S PEARS is the same, but without the green edge.

COUNTY FAIR (AVON), Colonial, Tidbit, $100.

:OUNTY FAIR (AVON), POMEGRANATE AND ;RAPES, Salad Plate, $30. Note the rare red edge. These plates are usually found with the green edge.

COUNTY FAIR (AVON), Colonial, Complete Set, $275.

PURPLE PLUMS (PETITE FRUITS), Snack Plate, $30.

"PETITE APPLES" (PETITE FRUITS), Colonial, Dinner Plate, $20.

"PETITE CHERRIES" (PETITE FRUITS), Snack Plate, $30. SPIDER-WEB (GREEN), Cup, $5.

FRUIT COCKTAIL SET, LEMONS, Astor, Salad Plate, $30.

Highly advertised fruit sets made by Southern Potteries (House Beautiful, 1942, 1949, and 1951).

FRUIT COCKTAIL SET, PINEAPPLE, Astor, Salad Plate, $50. This plate appears to be the most elusive.

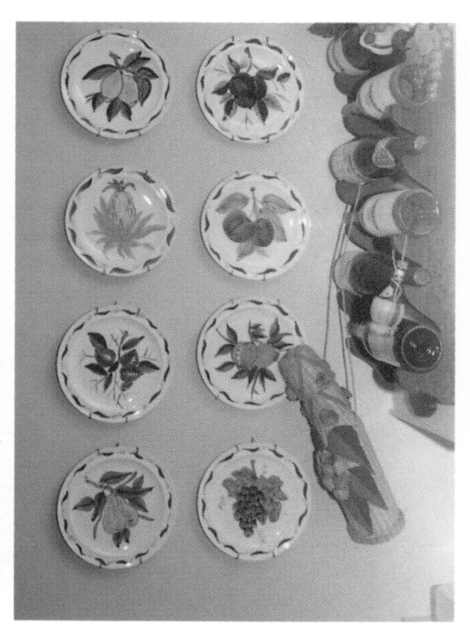

Complete set of FRUIT COCKTAIL, Astor, Salad Set.

DUFF Salad Set Plates, House Beautiful Magazine, September 1947.

UFF SALAD SET (ORANGE), with green swirl
ackground, $25. Background will vary.

"EGG CUSTARD" and "BUTTERMILK," Martha Pitchers, $65 each.

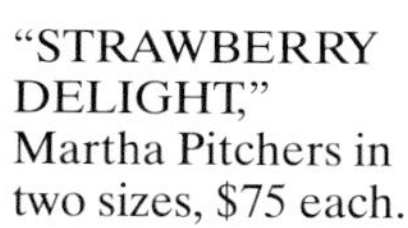

"STRAWBERRY DELIGHT," Martha Pitchers in two sizes, $75 each.

CHERRY ICE," Martha Pitcher, $75. "GUAVA,"
artha Pitcher, $100.

"KEY LIME," Martha Pitcher, $100. "LIME SHERBET,"
Spiral Pitcher, $75.

ARAMEL," Martha Pitcher, $100.

"GRAPE BLEND," Martha Pitcher, $65.

"BANANA PUDDING," Martha Pitcher, $65.

Vegetable Patterns

This line appears to have been produced almost exclusively for Vitamin Frolics. Collectors of this category wish more patterns could be found.

VEGGIE (CARROT AND SCALLIONS), Skyline, Snack Plate, $35.

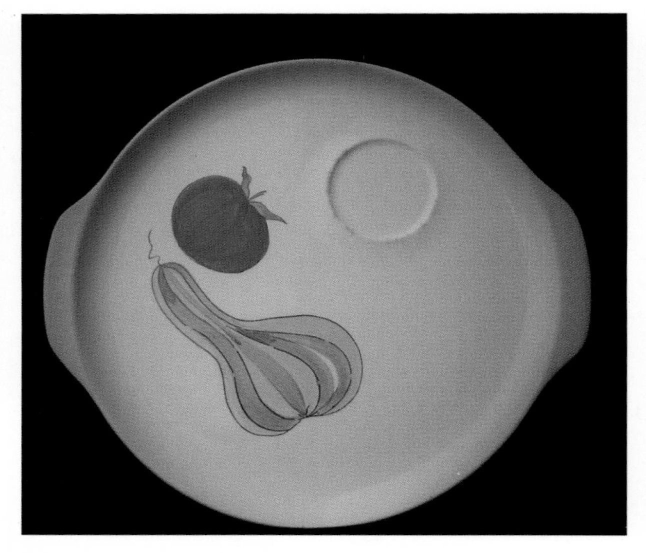

VEGGIE (TOMATO AND SQUASH), Skyline, Snack Plate, $35.

VEGGIE (TURNIPS AND PEAS), Skyline, Snack Plate, $35.

VEGGIE (BEET AND BEANS), Skyline, Snack Plate, $35.

VEGETABLE STEW," Salad Bowl, $75. There also a VEGETABLE SOUP salad bowl.

GREEN PEPPER, Candlewick, Bread and Butter Plate, $50. This is part of the VEGETABLE SOUP Set. Look for the green pepper in the salad bowl. These pieces have the Vitamin Frolics backstamp.

"RADISHES," VEGETABLE PATCH Series, Skyline, Tab Bowl, $25.

When this collector is not out finding Blue Ridge, she is ironing this extensive collection of 1950s tablecloths.

Chapt

Chapter Five
A Potpourri of Patterns

Solids, lines, and shape patterns were designed throughout the years of production at the pottery. Plaids and geometrics are representative of the later years. Border designs are the most popular designs in this area.

Garage collections are growing in favor as collectors run out of room in their homes. Many collectors store and sell the overflow of their collections from their garages.

SMOKY MOUNTAIN LAUREL, Candlewick, 12" Platter, $30. The simplicity of this pattern makes it a natural accent for many other patterns. Also found in a peach color.

This SMOKY MOUNTAIN LAUREL Plate has been "spiced up" with an artist's over-painting of two peaches.

"BLUE VICTORIAN," Martha Pitcher, $75.

"BLUE BEAUTY," Martha Pitchers in two sizes, $75 each.

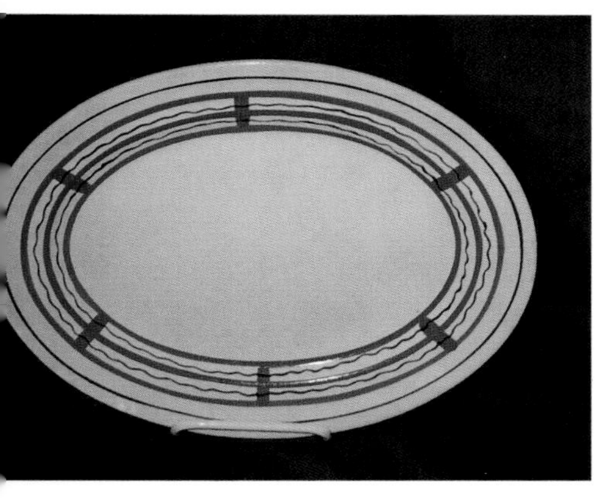

TRACKS," Platter, $45.

"CROSSBAR," Candlewick, Dinner Plate, $25.

Southern Potteries, Inc. label on the back of the "CROSSBAR" plate. Label indicated the Dec. No. (Decoration Number) as 2639U. The U usually indicates a variation of an existing pattern.

JUGGLING," Saucer, $4.

WEE LEAF, 9" Plate, $10.

AROUND ROSEY, Covered Casserole, $65.

"SPIRAL ME CREAM" and "SPIRAL ME BLUE," Spiral Pitchers, $100 each.

"BUCKY," Candlewick, Dinner Plate, $20.

"AQUA ABBY," Abby Jug, $95. Abby jugs are approximately 7" in height and have 6 circular rings.

SOUTHERN RUSTIC, Bread and Butter Plate, $8. This pattern is sometimes confused with ROUNDELAY, which has no break in its swirl.

"TEAL RUSTIC," Skyline, Teapot, $75. This resembles SOUTHERN RUSTIC but is in shades of black and teal.

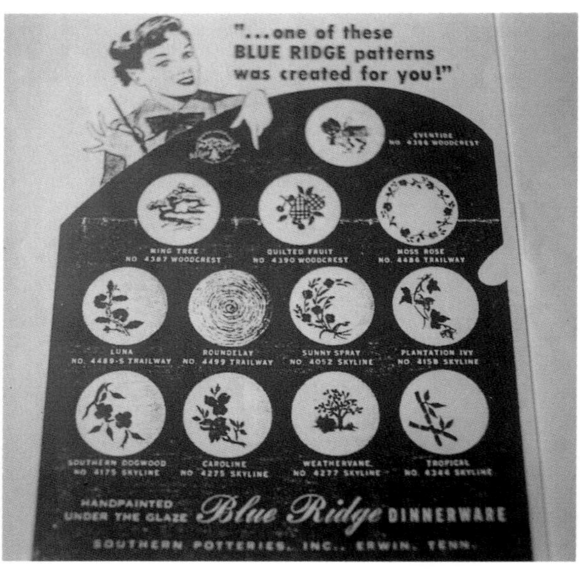

Advertisements were common during this period in Southern's production. This one (in BETTER HOMES AND GARDENS, July 1954) boasts "One of these Blue Ridge patterns was created for you!" Note the ROUNDELAY pieces.

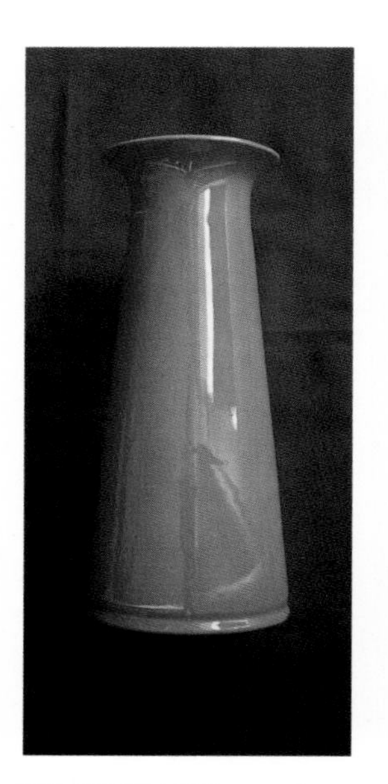

"SO BLUE," Tapered Vase, $100.

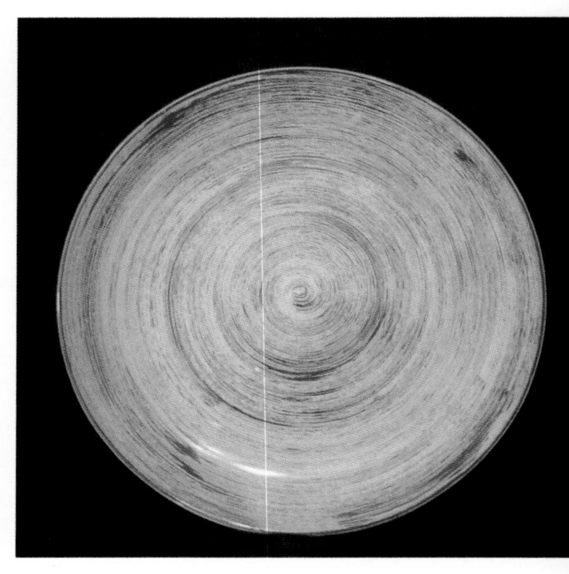

"ROUNDELAY (YELLOW)," Dinner Plate, $15. This pattern's variations in pink, blue, gray, and green are popular with admirers of 1950s china.

PECAN and SPIDERWEB Cups on Square Plate reflect the popularity of earth tones during the 1950s.

PECAN, Snack Plate and Cup, $25.

"DAY & NIGHT," Martha Pitcher, $95.

"SO PINK," Handled Vase, $100.

"MYSTIC MOSAIC (CORAL),"
Three-Tier Server, $40.

"MYSTIC MOSAIC (GREEN)," Salad Plate, $5.

"SPACE," Dinner Plate, $10. SPIDERWEB (BLACK), Creamer, $20, and Butter Dish, $35. Creamer is also used as a small pitcher.

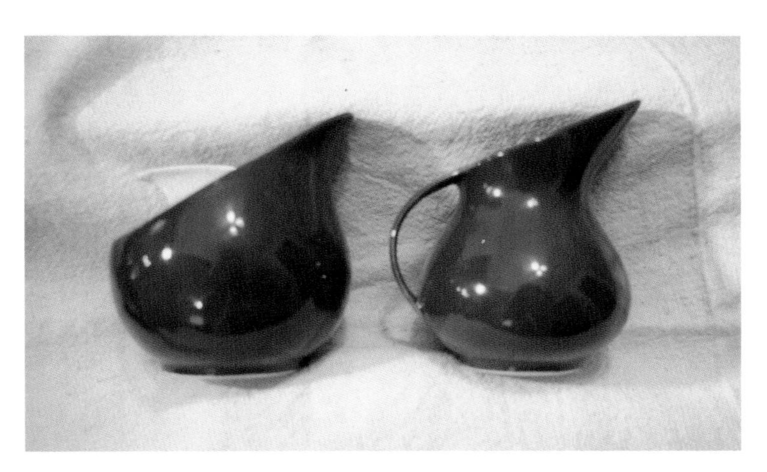

SPIDERWEB, Palisades, Sugar and Creamer Set, $45.

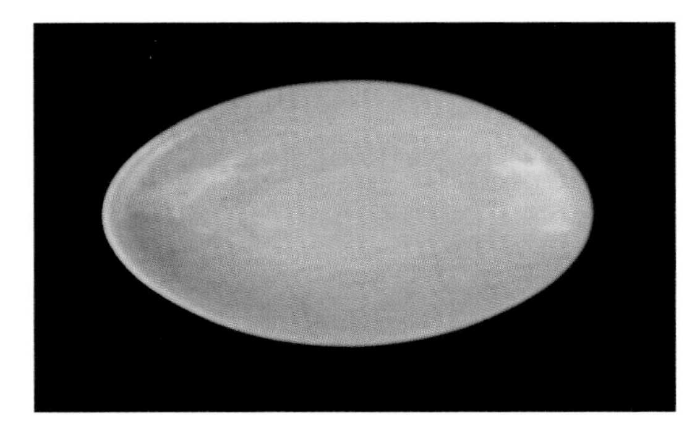

Oval Relish Dish, $15. These pink, yellow, and blue SPIDERWEB pieces are sometimes called MARBLE.

SPIDERWEB, Ovide, Coffee Pot, $125. Note how lid was made to fit down into the pot.

SPIDERWEB (GRAY), Skyline, Butter Dish, $35. Salt and Pepper Shakers, $20. Palisades Pitcher, $45.

Display case showing sets of (Top to bottom) MAYFLOWER, RUSTIC PLAID, and SARASOTA.

SPIDERWEB (BLACK), Skyline, Teapot, $65. Lids come in an assortment of colors.

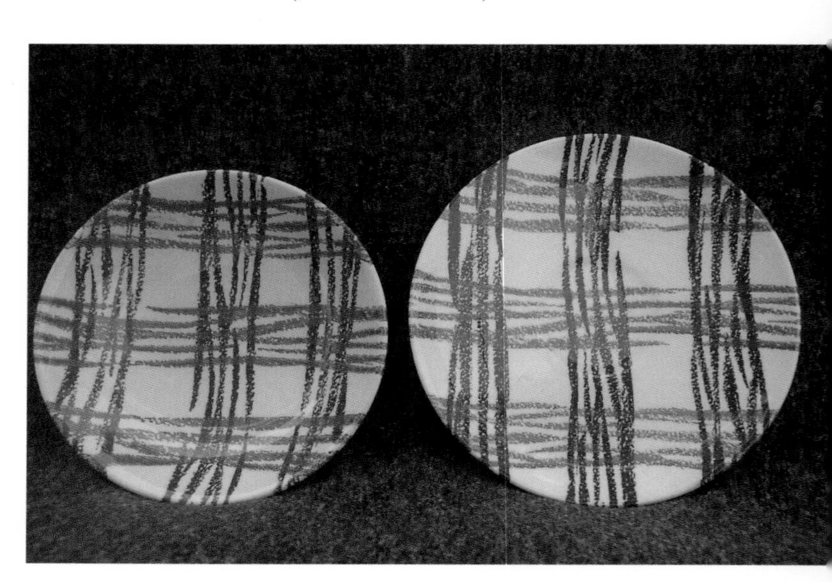

ROCKY PLAID, Saucer, $3. Bread and Butter Plate, $5.

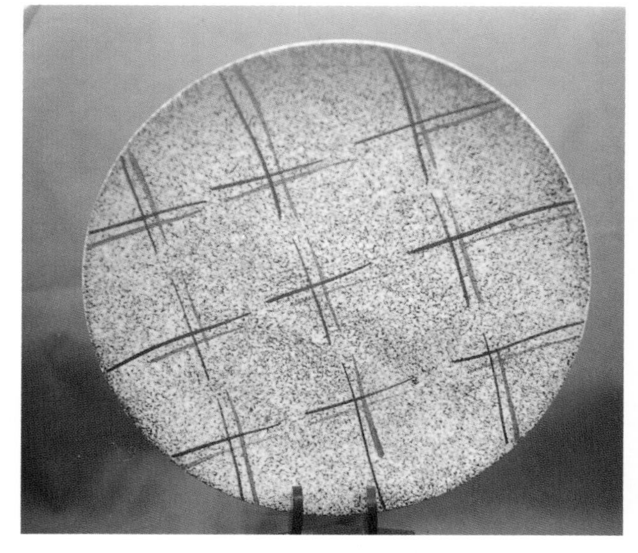

LORETTA, Dinner Plate, $5. DUSTY and JOHN'S PLAID are color variations of this pattern.

JOHN'S PLAID and "GREEN PLAID," Skyline, Teapots, $65 each.

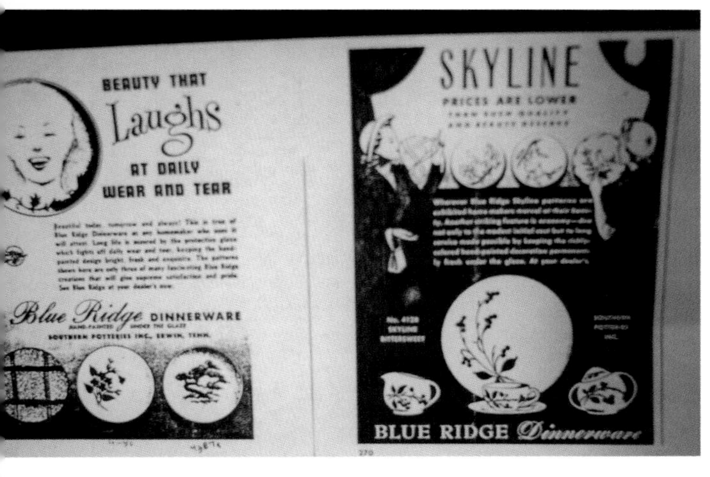

USTIC PLAID, FRENCH VIOLETS, and MING TREE are beauty that laughs at daily wear and tear" (HOUSE BEAUTIFUL, 954). BITTERSWEET was introduced when "prices are lower than ver" (BETTER HOMES AND GARDENS, 1951).

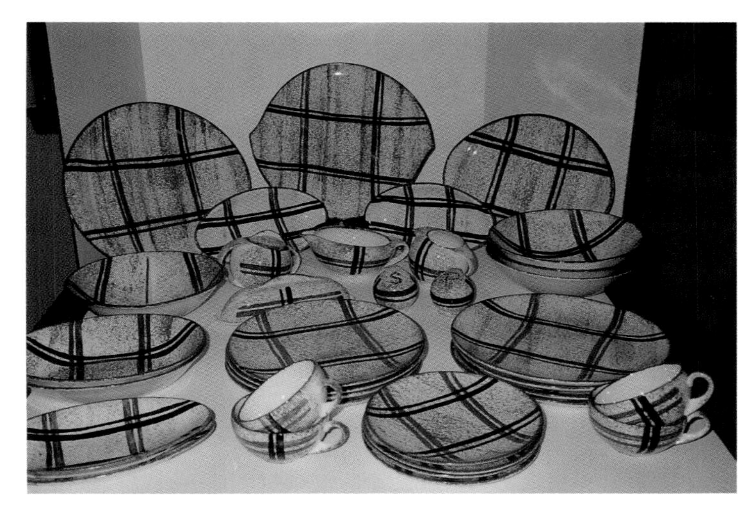

Set of "MAROON PLAID." Note Palisades platter.

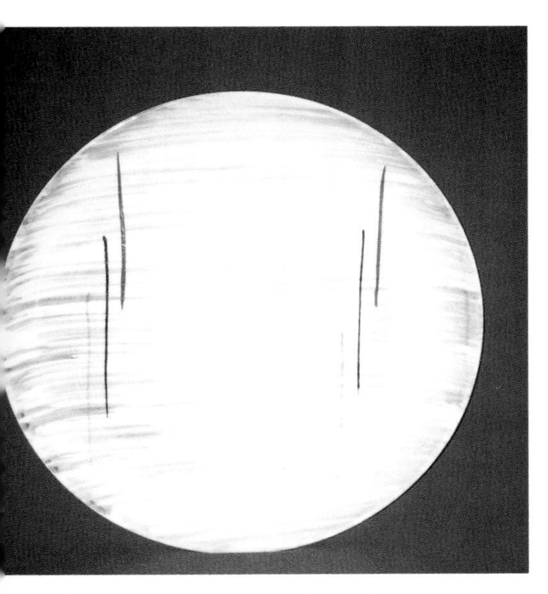

RAINY DAY, Dinner Plate, $10. Primrose China backstamp.

FRIENDSHIP PLAID, Bread and Butter Plate, $3. This pattern is very similar to "WAVY PLAID."

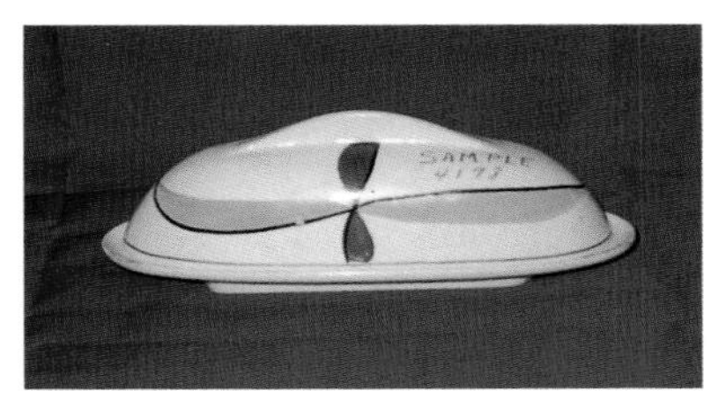

"TWISTED RIBBONS," Butter Dish, $100. Note the pattern number, 4178.

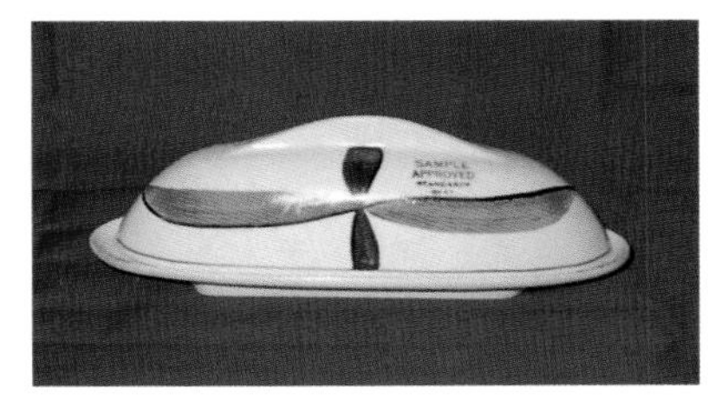

Other side of "TWISTED RIBBONS" Butter Dish, indicating SAMPLE APPROVED, STANDARDS DEPARTMENT.

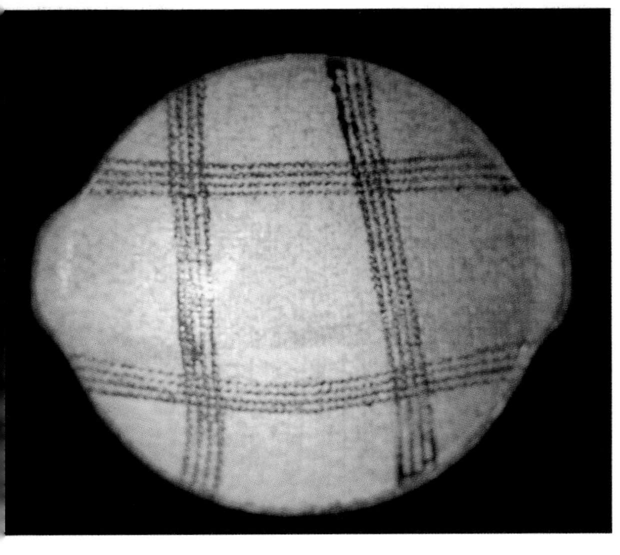

STENCIL (GREEN), Woodcrest, Tab Bowl, $20. Also found in brown and gray.

"TERESIA'S JUG," Abby Jug, $75. Most Abby jugs are unmarked. The key is the applied handle.

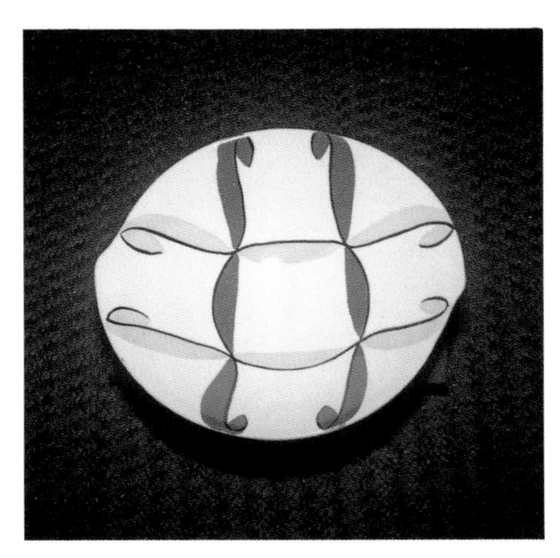

RIBBON PLAID, Palisades, Soup Bowl, $10.

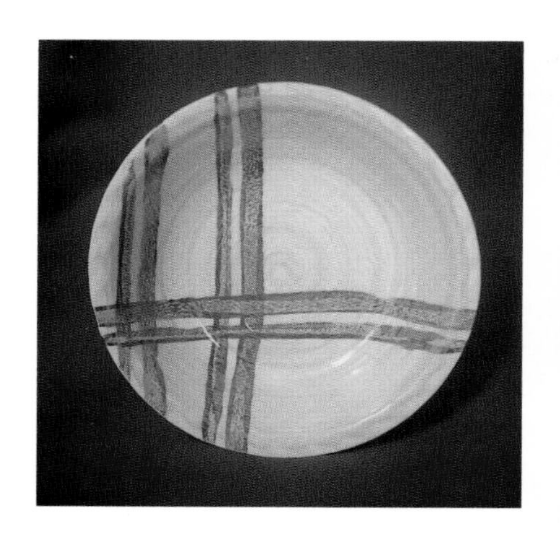

PIEDMONT PLAID (GREEN), Piecrust, Fruit Bowl, $5.

Set of "LIME ICE" accented with Cash pitchers and Erwin Pottery shakers.

SANDRA, Skyline Teapot, $95.

"LIMEY," Abby Jug, $50. Unmarked but generally attributed to Southern Potteries early production.

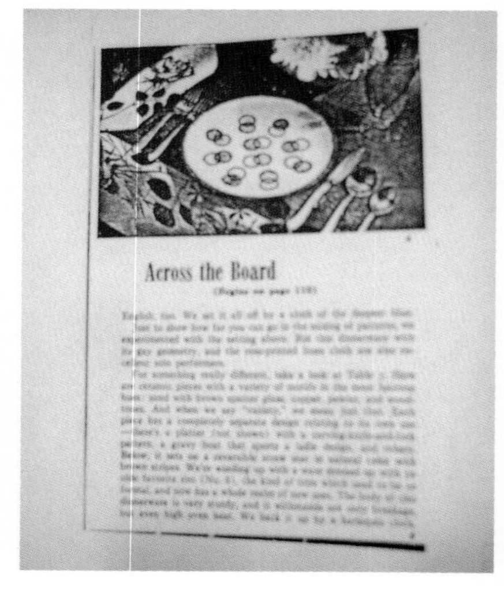

Across the Board

BUBBLES is used to illustrate variety in table arrangements in this unidentified article from 1950.

"SEA GREEN," Martha Pitchers in two sizes, $75 each.

"LINDA," Skyline, Creamer and Sugar with lid, $25 set.

LEAF & CIRCLE, Skyline, 10" Plate, $25. Look for rope handle accessories.

"BOOMERANG," Divided Dish, $35. A very similar pattern in blue is called ROBIN.

"BANANAS & CREAM," Ovide, Coffee Pot, $100.

"FLOWER TREE," Demi Cup and Saucer, $50. QUARTET, 4" Square Tile, $25. QUARTET is also the name of a CHERRY pattern.

"WATERMELON SLICE," Dinner Plate, $15.

"BORDER TEAL LEAF," 8.75" Square Plate, $20. Teal line at the edge is unusual.

LEAF, 9.5" Square Plate, $30. Unusual size. ANTIQUE LEAF, Lacy Scroll Edge, Tab Bowl, $25. Marked "OVENPROOF."

PASTEL LEAF, Candlewick, 8" Vegetable Bowl, $15. There is also another pattern called "PASTEL LEAF."

"LYDIA," Candlewick, Dinner Plate, $20.

LEAF, Creamer, $25. This unusual creamer with the "eared" handle has been identified as an early piece.

"RINGED LEAF," Big Cup and Saucer, $65.

MADRAS, 7" Square Plate, $25.

Chapter Six
Special Patterns

"MARINER VARIANT," Coaster/Butter Pat, $100.

(Top to Bottom) "CRUISING," 9" Plate, $125, SAILBOAT, 6" Plate, $100, and MARINER, Coaster, $100.

TUNA SALAD, 7" Plate on Wooden Stand, $45.

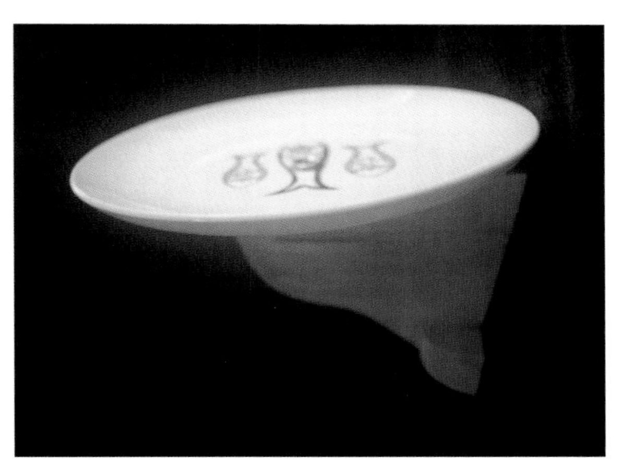

Side view of unique wooden stand.

Designer

"PARABOLA," DESIGNER SERIES, Clinchfield, $200.

Back of "PARABOLA" plate, showing the initials PGH.

GLIMMER, DESIGNER SERIES, Cake Plate, $100. Also found on Skyline shape. Most Designer patterns are on Clinchfield shape.

You won't find canned goods in this pantry, but you will find two designer plates, ABRACADABRA and "FEATHERS" on the top shelf.

Feathered Friends

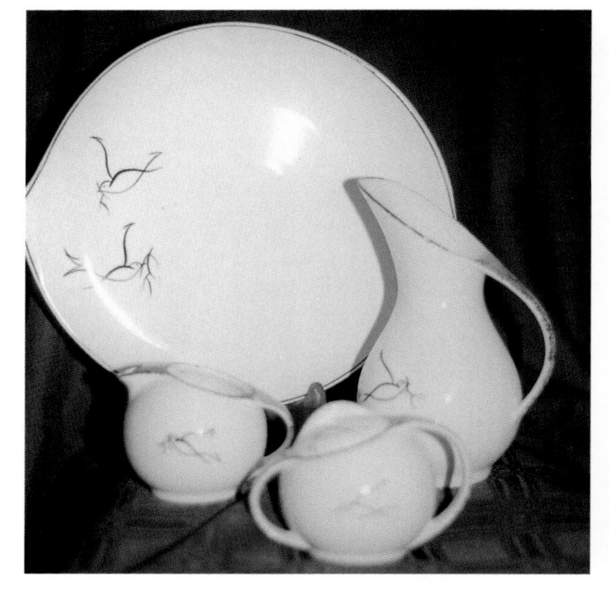

NESTING BIRDS, Palisades, Platter, $50, Palisades, Pitcher, $65, NESTING BIRDS, Skyline, Creamer and Sugar, set, $60.

TWEET, Astor, 8" Bowl, $45. GREEN FEATHER is also a single bird on the identical shape.

tamp used for DISTLEFINK pattern.

DISTLEFINK, Plate, $25.

)aniel Miller, great-grandson of Novella
ealls, with "BROWN THRASHER" from
e Astor SONGBIRDS SET.

BLACKBIRDS, Flat Soup Bowl, $45.

"RUSTY BLACKBIRD" (SONGBIRDS), Astor, Salad Plate, $100. Tree and leaf placement may vary.

"GRAY CATBIRD," (SONGBIRDS), Astor, Salad Plate, $95.

HERMIT THRUSH (SONGBIRDS), Astor, Salad Plate, $95.

FINCH, (COLONIAL BIRDS SET), Dinner Plate, $100.

ORCHARD ORIOLE (COLONIAL BIRDS SET), 8.5" Plate, $100. If you look closely at this painting, you can see that the misplacement of the branch makes it appear that the bird has a worm in its mouth.

WESTERN BLUEBIRD, Colonial, Dinner Plate, $100.

LOUISIANNA WOODTHRUSH (COLONIAL BIRDS SET), Dinner Plate, $100.

TUFTED TITMOUSE, Colonial, Dinner Plate, $100. Sometimes lighter in color.

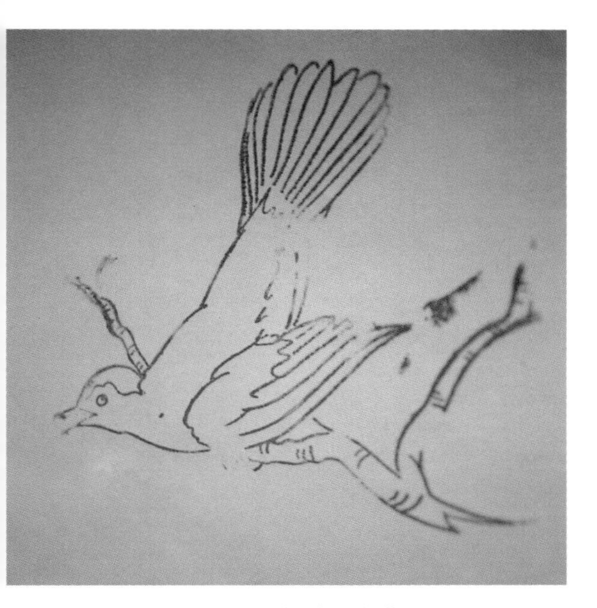

Stamp used for the catbird in the Skyline SONGBIRDS SET.

Complete set of Skyline SONGBIRDS. This has been identified as a luncheon set, but to date, no cups and saucers have been found. Perhaps the plates were meant to be decorative.

PEACOCK, Candy Box, $300. The tulips on the bottom are subtly repeated on the top.

REDBIRD, (SONGBIRDS), Skyline, Salad Plate, $75.

DREAMBIRDS, Platter, $125.

RUBY-THROATED HUMMINGBIRD (SONGBIRDS), Skyline, Luncheon Plate, $85.

Table Set with DREAMBIRDS.

YORKTOWN, Colonial, Individual Server, $125. YORKTOWN is one of the TALISMAN wallpaper patterns.

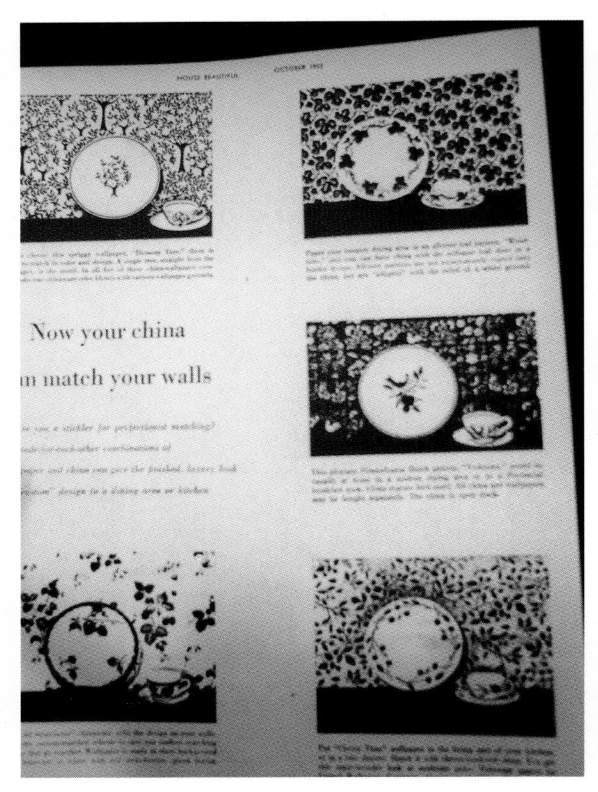

Now your china

n match your walls

United Wallpaper Company's Talisman line boasted "Now your china can match your walls" with the following patterns: BLOSSOM TREE, WOODBINE, YORKTIME, WILD STRAWBERRY, and CHERRY TIME.

Caribbean Series

SAILFISH (CARIBBEAN SERIES), Plate, $500.

FLAMINGO POND (CARIBBEAN SERIES), 8.5" Plate, $500.

PARROT JUNGLE (CARIBBEAN SERIES), Astor, Plate, $800. Note the teeth.

CARILLON (CARIBBEAN SERIES), Plate, $800. This plate is extremely rare.

rtist-Signed Pieces

UAIL, Clinchfield, 17" Unsigned Platter, $1200.

QUAIL, Colonial, 12" Plate, $800. Signed Mildred Broyles.

QUAIL, Colonial, 12" Plate, $800. Signed Nelsene Quesenbury. Most of this artist's pieces are signed Nelsene Q. Calhoun or Nelsene Calhoun.

QUAIL, 12" Plate, $800. Signed Nelsene Calhoun.

QUAIL, Colonial, 12" Colonial Platter, $1,000. Although unsigned, this is an extremely rare piece.

TURKEY GOBBLER, Clinchfield, 17.5" Platter, $1,200.
Signed Mae Garland.

TURKEY GOBBLER, Clinchfield, Platter, $1,200.
Signed Frances Kyker.

Signature of Mae Garland.

TURKEY GOBBLER, Saucer, $500.
Signed Alleene Miller.

TURKEY GOBBLER, Saucers. Top signed by Nelsene Calhoun, Bottom left signed by Mae Garland, Bottom right signed by Alleene Miller. Each $500.

WILD TURKEY, Clinchfield, 19" Platter, $1,500. Signed Ruby S. Hart. Previously called TURKEY HEN, the name for this platter was changed after close observation of an Audubon drawing proved the hen's true gender!

FLOWER CABIN, Clinchfield, 10" Plate, $800. Signed Mae Garland.

GOLD CABIN, Clinchfield, 10" Plate, $800. Signed Louise Guinn.

FLOWER CABIN, Clinchfield, 10" Plate, $800. Signed Mae G. Hice. This plate was also painted by Mae Garland but after her marriage.

GOLD CABIN, Clinchfield, 10" Plate, $800. Signed Nelsene Q. Calhoun.

FLOWER CABIN, Clinchfield, Plate, $800. Signed Frances Kyker. Every artist added a personal touch.

Character Jugs

PIONEER WOMAN, Character Jug, $800.

The PIONEER WOMAN with a carrot top, surrounded by delicious looking meals in this page from the *BETTY CROCKER NEW PICTURE COOKBOOK*, 1961, McGraw Hill, Company.

Incised bottom of the PIONEER WOMAN.

Side view of the PIONEER WOMAN.

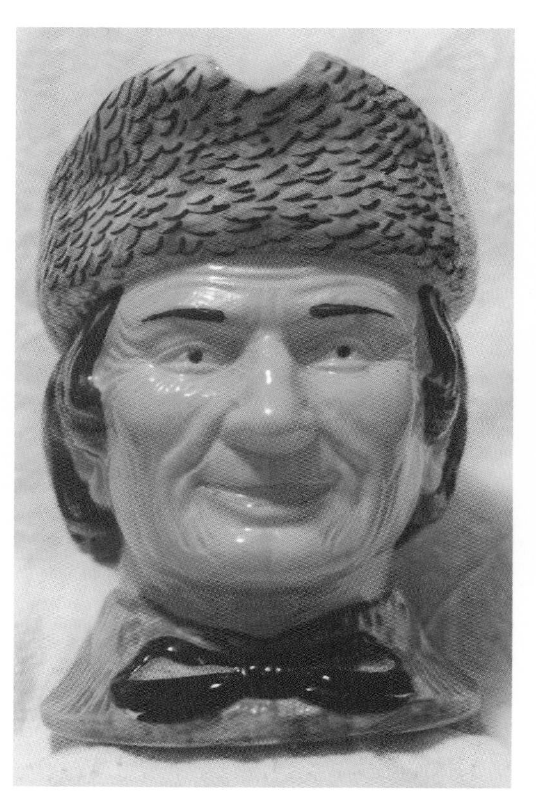

DANIEL BOONE, Character Jug, $800.

Close up of the handle on the DANIEL BOONE Jug.

Side view of DANIEL BOONE.

Incised bottom of DANIEL BOONE Jug with Blue Ridge backstamp.

Hand-carved sculpture from which the DANIEL BOONE mold was made.

Smaller hand-sculpted DANIEL BOONE head.

PAUL REVERE, Character Jug, $800.

PAUL REVERE sculpture used to make the mold for REVERE character jug.

Incised bottom of PAUL REVERE Jug.

Side view of PAUL REVERE.

Character jugs are quite desirable pieces for book collectors. They make wonderful, but expensive, book ends.

THE INDIAN, Character Jug, $950. The rarest of the set.

Country Life Series

JENNY WREN (COUNTRY LIFE SERIES), 6" Square Plate, $150.

CRADLE (COUNTRY LIFE SERIES), 6" Square Plate, $200. Others in the series include CHURN, GRANDFATHER CLOCK, SPINNING WHEEL, and FISHERMAN.

This lovely quilt depicting Blue Ridge patterns was a gift from Yvonne Waggoner to Judy Leach.

Countryside Series

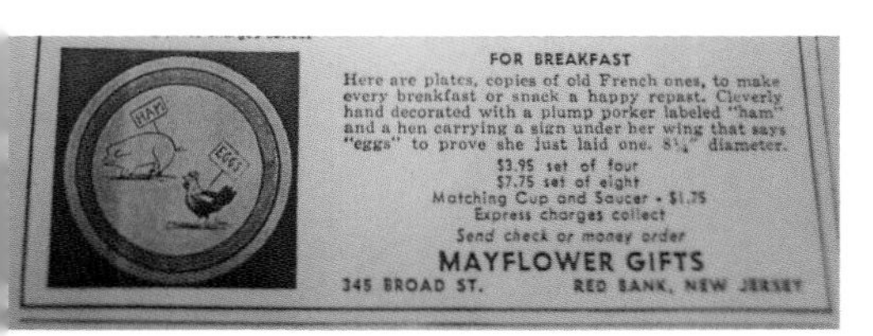

FOR BREAKFAST

Here are plates, copies of old French ones, to make every breakfast or snack a happy repast. Cleverly hand decorated with a plump porker labeled "ham" and a hen carrying a sign under her wing that says "eggs" to prove she just laid one. 8¼" diameter.

$3.95 set of four
$7.75 set of eight
Matching Cup and Saucer • $1.75
Express charges collect
Send check or money order

MAYFLOWER GIFTS
345 BROAD ST. RED BANK, NEW JERSEY

Mayflower Gifts, New Jersey, advertises the HAM AND EGGS pattern: "Here are plates, copies of old French ones, to make every breakfast or snack a happy repast. Cleverly hand decorated with a plump porker labeled "ham" and a hen carrying a sign under her wing that says "eggs" to prove she just laid one. 8.25 diameter. Prices: $3.95 for a set of four. $7.75 for a set of eight. Matching cup and saucer for $1.75." House Beautiful, March 1940.

HAM 'N EGGS for breakfast!

SHELLING PEAS (COUNTRYSIDE SERIES), Candlewick, Flat Soup Bowl, $50. This series is sometimes referred to as RURAL LIVING.

HAM 'N EGGS, Candlewick, 12" Plate, $75. Used as underplate for matching salad bowl. Also referred to as a salver.

Sugar and Creamer, Candlewick, Countryside Motif, Unmarked. These are thought to be from the Countryside Series but have not been authenticated.

(Left to right) STRAW HAT (COUNTRYSIDE SERIES), Candlewick, Saucer, $35. "SUN HAT" (COUNTRYSIDE SERIES), Candlewick, Saucer/Coaster, $35.

"DOTTED CLUCKERS," Salt and Pepper Shakers were made by Mabel Duncan Cash while she worked at Southern. They have her initials on the bottom under the glaze.

CHANTICLEER, Skyline, Divided Vegetable Bowl, $75.

COCK-A-DOODLE, Skyline, 3-Tier Tidbit, $100.

"ARAB," Chick Pitcher, $175.

ROOSTER MOTTO, 6" Square Plate, $300. Although this is a rare piece, the ROOSTER MOTTO plate is also found on Candlewick. Some have the PV backstamp while others are unmarked.

WEATHERVANE COCK, Square Plate, $65. Similar plates have been found marked: "Made in France."

COCKY-LOCKY, 12" Candlewick, Cake Plate, $125.

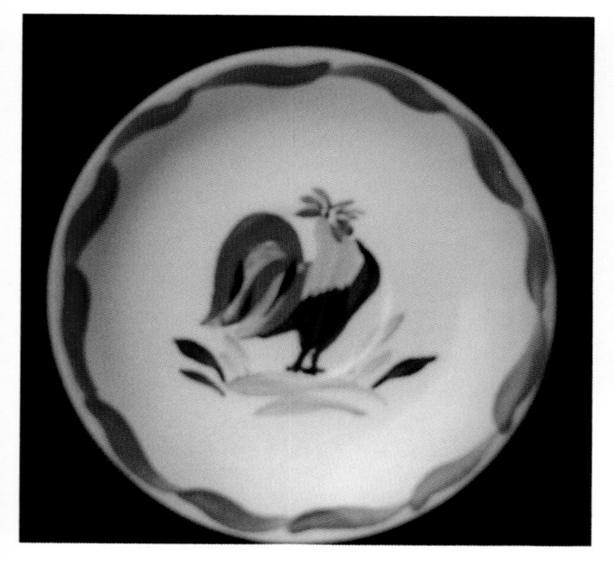

Above: GAMECOCK, Clinchfield, Bread and Butter Plate, $70, Fruit Bowl, $50.

Left: COCKY-LOCKY, Clinchfield, Saucer, $65.

Italian Plate: "MY LOVE WILL STOP WHEN THIS ROOSTER CROWS." Note feet and tail feathers.

"STREAKERS," Salt and Pepper Shakers, $125 pair. Also recently spotted – a white chicken with polka dots.

"CLUCKERS" and "DOMINICKERS" are a favorite in country kitchens, $125 each pair. Sculptured fruit pitchers are found in three sizes, $100 each. FALLING LEAVES, Colonial, Dinner Plate, $25.

EVENTIDE, Woodcrest, Dinner Plate, $40, Teapot, $175.

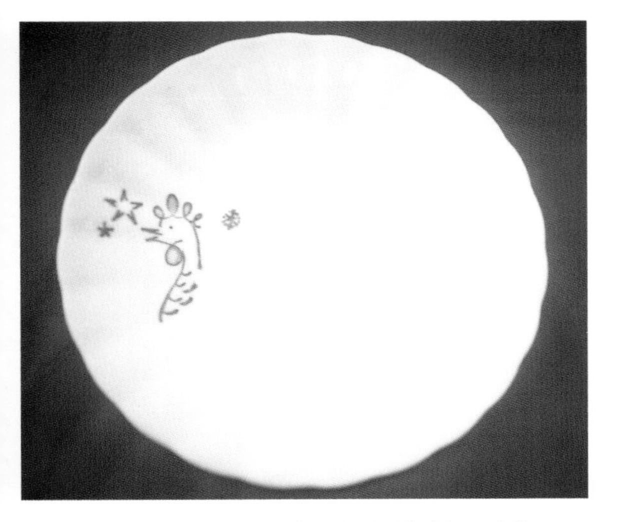

ROCKPORT ROOSTER (PROFILE), Plate, $50. The full rooster can be found on the Skyline shape.

A bove: RED ROOSTER Plate and Southern Potteries, Inc. Shipping Box stamped: "53 PC SET, DEC 4422, DISMAYED ROOSTER."

Left: BLUE LINE FARM, 9" Bowl, $60. COCK-A-DOODLE, 9" Bowl, $50.

FARMYARD, Skyline, 9" Plate, $125.

Blue Ridge egg plate on left and English one on the right. Both have PV backstamps.

Stamp used for barn and silo in BELLE HAVEN.

WEATHER VANE was presented as "BREEZY AND GAY" in this magazine ad. Also featured were CALICO FARM, TROPICAL, FRIENDSHIP (4321X), and CAROLINE.

WEATHERVANE, Maple Leaf Tray, $125.

BELLE HAVEN, In the chicken coop, Plates, $100 each.

ROOSTER, Egg Plate, $100. Shouldn't this be called "hen" egg plate?

HILLSIDE, Piecrust, 9" Plate, $150.

Back of HILLSIDE Plate, initials ARM under the glaze, thought to be Alleene Miller.

HOMEPLACE, Skyline, Sugar, $55 with lid. 10" Plate, $125.

Candy Ellison and daughter Laura show off their favorite patterns, FLOWER CABIN and HILLSIDE.

FOLKLORE, Skyline, 10" Plate, $150. Also found on Woodcrest shape.

SONORA, 7" Plate, $250. Extremely rare pattern.

HOMESTEAD, Skyline, Dinner Plate, $450.

Back of HOMESTEAD Plate, designating pattern as COUNTRYSIDE #1.

"LAZY DAY FARMS," Dinner Plate, $250.

Back of "LAZY DAY FARMS" Dinner Plate, designating the plate as Tem. #343.

"ROPE 'EM, COWBOY," Dinner Plate, $250.

Back of "ROPE 'EM, COWBOY" Dinner Plate. Signed "LEN – 15."

SQUARE DANCE, Piecrust, Salad Bowl, $125. Look for a carafe, too.

"THE CALLER," SQUARE DANCE, Square Round Plate, $250. It was previously presumed that there were only eight SQUARE DANCE plates. "Ready? Tap your feet!"

MEXICANO, Astor, Saucer, $60.

MEXICANO, Astor, Creamer, $75. On some pieces of this pattern, Senor is taking a siesta.

SQUARE DANCE, #1 "Start the music." #2 "Choose your partner."

SQUARE DANCE, #3 "Greet your partner." #4 "Circle around."

SQUARE DANCE, #5 "Hug your partner." #6 "Swing your partner."

SQUARE DANCE, #7 "Step in line." #8 "Promenade home."

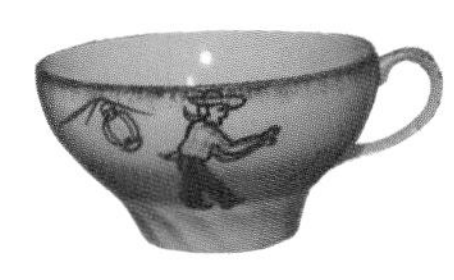

You'll need this SQUARE DANCE
Cup after all that dancing!

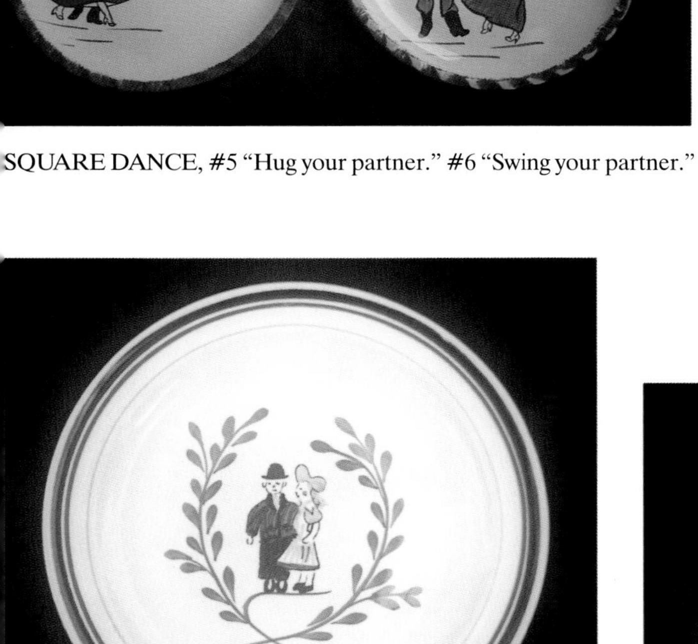

LOVE SONG, Astor, 8.5" Plate, $350. A special pattern. Reportedly, designed and manufactured as favors for wedding attendants. Only plates have been found.

DO-SI-DO, Clinchfield, 11.5" Platter, $300. A rare pattern, but not a specialty item. Dinnerware pieces are available, but expensive.

Provincial Line

MAN WITH PITCHFORK, 5" Square Plate, $65. MOWING HAY, 5" Square Plate, $75. These are part of the PROVINCIAL FARM SET made for and backstamped PV. The red border seems to be more difficult to find.

MOWING HAY, 11.5" Cake Plate, $125. This pattern is from the PROVINCIAL FARM SET.

BRITTANY, Partial Breakfast Set, $750.

SWISS DANCERS, Flat Shell Bon-Bon, $275.

BRITTANY, Candlewick, 11.5" Plate, $75.

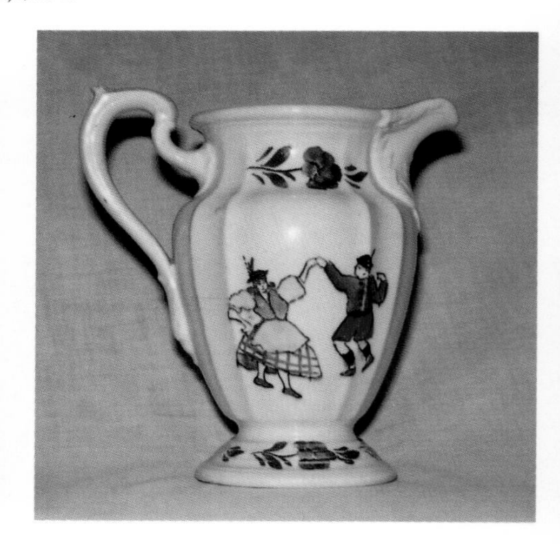

SWISS DANCERS, Milady Pitcher, $300.

Above: LYONNAISE, Square Box, $175.

Left: LYONNAISE, Candlewick, 11.5" Cake Plates, $125 each.

NORMANDY, Skyline, Square Plate, $45. Pepper Shaker, $65 set.

LYONNAISE, Clinchfield, Platter, $200. Dinner Plates, $125 each. Bread and Butter Plates, $45 each. Covered Sugar, $65. LYONNAISE is becoming more and more desirable to collectors. It is approaching French Peasant status.

LYONNAISE, Unmarked Cups, without the line treatment.

LYONNAISE (MADEMOISELLE), Clinchfield, Dinner Plate, $125. LYONNAISE (MONSIEUR), Luncheon Plate, $125.

LE COQ SOLEIL, Astor, Platter, $175. LE COQ, Clinchfield, Plate, $100.

French Peasant pieces with the rose color leaf.

CALAIS, Dinner Plate, $125. Also found on Clinchfield shape with wide pink ring around inner edge.

FRENCH PEASANT, Chocolate Set, $1300.

FRENCH PEASANT, Spiral Pitcher, $150.

FRENCH PEASANT, Antique Pitcher, $200.

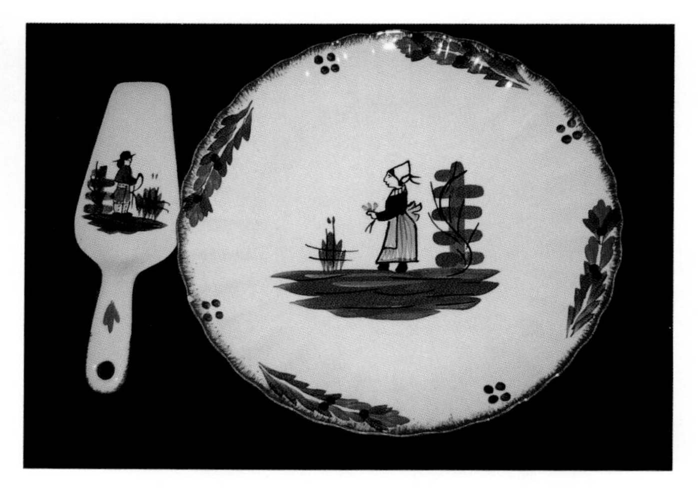

FRENCH PEASANT, Cake Lifter and Cake Plate, $250.

FRENCH PEASANT, Bottom of Covered Casserole, $100. Note the ruffled edge of the bowl.

FRENCH PEASANT, Salad Bowl with Underplate, $350.
Pieces have persimmon color leaf.

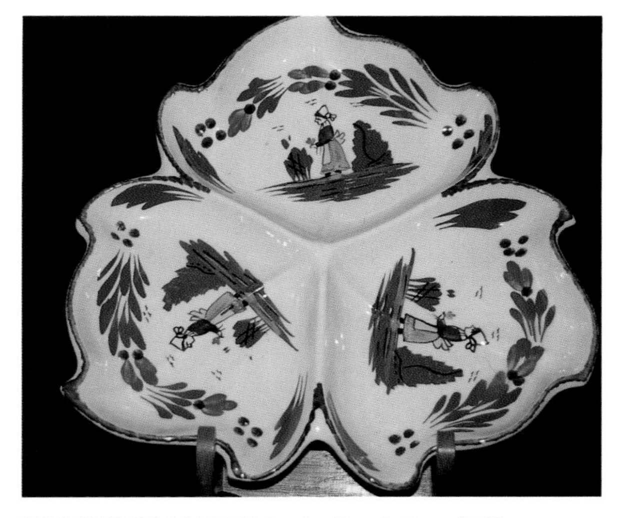

FRENCH PEASANT, Martha Snack Tray, $300.

FRENCH PEASANT, Ruffle Top
Vase, $250. Both sides shown.

Chapter Seven
Seasonal Patterns

The imagination of the designers created an impressive array of patterns suitable for celebrating the seasons and special days of our lives.

Beautiful Thanksgiving table set with TURKEY WITH ACORNS.

PILGRIMS, Skyline, Dinner Plate, $60.

TURKEY WITH ACORNS, Skyline, Platter, $325. Cups and saucers have no turkey, only leaves and acorns.

THANKSGIVING TURKEY, Clinchfield, Dinner Plate, $100. MALLARD SHAKERS, $500. The set of shakers shown here was made in 1950 as a special wedding present.

RED FLOWER BETSY seems to be eyeing the "HAYSTACK" cups and saucer seen here.

TURKEY WITH ACORNS Plate has a prominent place in this display.

MOD OAK LEAF, Piecrust, Square Round Plate, $25.

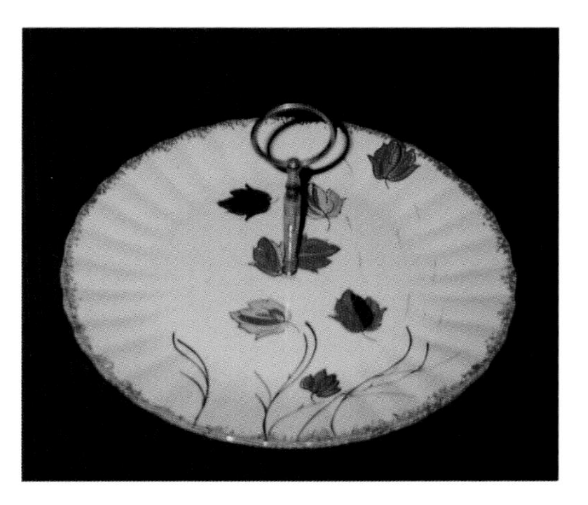

FALLING LEAVES, Tidbit, $40. Thanksgiving patterns have not been found on serving pieces. FALL COLORS and FALLING LEAVES are perfect complimentary patterns for this colorful season.

This appears to be a THANKSGIVING TURKEY dinner plate. However, it is backstamped "Rhythm" by Homer Laughlin. We believe it is a practice plate or a personal piece. It is inexplicable but charming.

TURKEY WITH ACORNS and THANKSGIVING TURKEY Plates await the big turkey day feast. Another Thanksgiving pattern (not shown) is TURKEY SURPRISE, Skyline, which has a wide yellow border.

PETAL POINT and Christmas go together.

FALL COLORS, Grace Pitcher, $150.

POINSETTIA continues to be "luxury with economy." (BETTER HOMES & GARDENS, January 1950) Although very popular today, pieces are readily available because of the large numbers of wares originally produced.

Water glass in POINSETTIA pattern. These were advertised in several catalogs. Maker was unidentified.

PETAL POINT shown in the Spring (1952) Montgomery Ward catalog (Chicago.)

POINSETTIA, Tidbit Server, $45.

Glasses in two sizes to go with PETAL POINT.

POINSETTIA, Colonial, Teapot, $160.

PETAL POINT, 2-Tier Server with Original Wooden Handle, $65.

HOLLYBERRY, Cup, $35. CHRISTMAS DOORWAY, Dinner Plate, $125. CHRISTMAS ORNAMENT, Saucer, $10.

CHRISTMAS ROSE, Luncheon Plate, $20. The # marks distinguish this pattern from WINK.

CHRISTMAS DOORWAY, Dinner Plate, $125. Cups have a wreath on one side and a Christmas tree on the other. They are difficult to find.

Blue Ridge collector's Christmas card picture using CLOVER WREATH with CHRISTMAS DOORWAY and CHRISTMAS TREE WITH MISTLETOE. CHRISTMAS TREE Skyline snack sets were boxed and sold as a "Party Time Set" in the early 1950s.

"POINTER," Skyline, Dinner Plate, $25.

CANTATA, 14" Platter, $45.

WINTERTIME, Divided Vegetable Bowl, $60.

A Blue Ridge Christmas!

Chapter Eight
Cookware

The demand for matching baking dishes and serving pieces resulted in a substantial number of these items being produced by Southern. The condition of pieces today attests to the fact that, although they were well-made, they were also well-used.

"BLUE PLUMES," Martha, Mixing Bowls, Set of three, $125. Sizes are 4.75", 5.75", and 6.75".

Literally, a Blue Ridge cake lifter rack. (Left to Right) LEAF, BORDER PRINT, HAPPY HOME, TULIP, DAISY CHAIN, DOUBLE DUTCH, BRIDESMAID or VICTORIA, FRUIT FANTASY or FRUIT PUNCH, CRAB or RED APPLE, and LEAF (color variation).

"NASCO ROSES," Cake Lifter, $25.

"SWEET BOUQUET," Round Divided Baking Dish, $45.

SPINDRIFT, Medium-size Leftover Container with lid, $40. Interesting interpretation of leaves on bottom.

DAISY CHAIN, Mixing Bowl, $50.

"NASCO ROSES," Divided Baking Dish, $45.

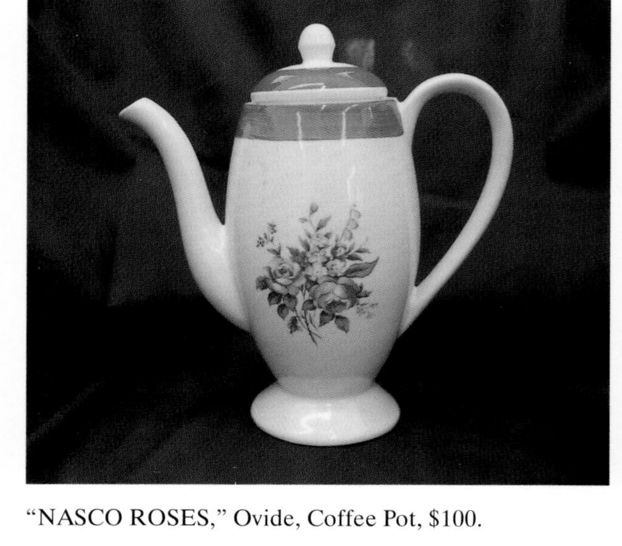

"NASCO ROSES," Ovide, Coffee Pot, $100.

MARDI GRAS, Range Shakers, $60.

YELLOW NOCTURNE, Ramikin, $40. The same mold was used to make the ramikin, the lazy susan center bowl, and the covered powder box.

RED TULIP, Covered Casserole, $75. This is a heavy Baking piece. It measures approximately 8" across and 4" deep.

CROSS STITCH, Divided Baking Dish with Stand, $45. Cake Lifter, $25.

GRANDMOTHERS GARDEN, Four-Section Warming Dish, $65

LEAF AND BAR, Five-Section Warming Dish, $55.

Royal backstamp on LEAF AND BAR cookware. Many other cookware patterns will be found with this backstamp.

ROCK ROSE, Four-Section Warming Dish, $55.

LEAF & BAR, 13" Oval Baking Dish in Aluminum Holder, $65. This piece has the Royal backstamp.

ROCK ROSE, Baking Dish, $45.

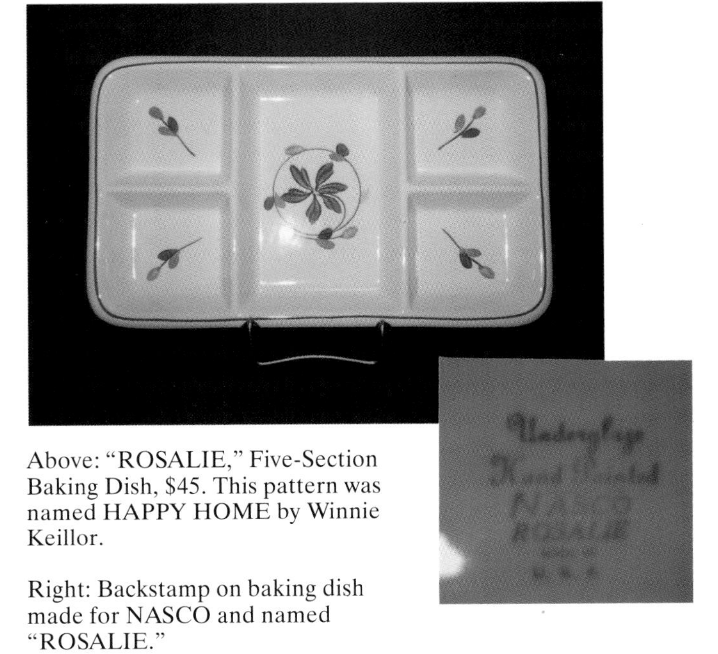

Above: "ROSALIE," Five-Section Baking Dish, $45. This pattern was named HAPPY HOME by Winnie Keillor.

Right: Backstamp on baking dish made for NASCO and named "ROSALIE."

BUTTERFLY AND LEAVES, Divided Baking Dish, $55.

MARINER, Set of Three Leftover Containers with Lids, $500.

LEAF, Trellis, Early Pitcher, $80

Geneva Kitchens often used Blue Ridge pieces in their advertisements. (House Beautiful, October 1948)

Above: POND IRIS, Mixing Bowl, $40.

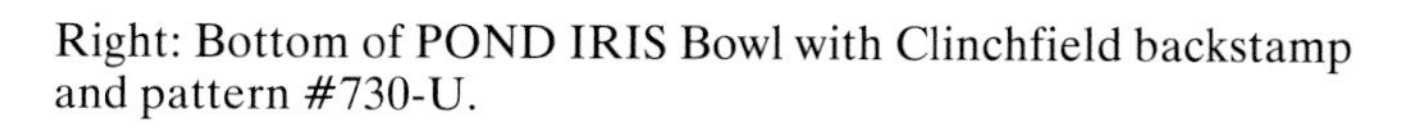

Right: Bottom of POND IRIS Bowl with Clinchfield backstamp and pattern #730-U.

VINCA, Skillet with lid, $80. Also called LATTICE FLOWER.

COREOPSIS, Round Divided Warming Dish, $65.

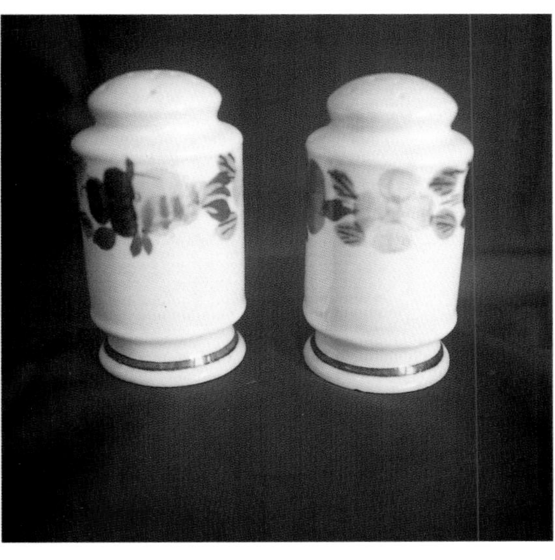

MARY, Range Shakers, $70. Range shakers were made in many patterns.

BORDER PRINT, Small Mixing Bowl, $20.

MARY, Divided Warming Dish, $95.

WILD IRISH ROSE (YELLOW), Round Divided Warming Dish, $65.

Chapter Nine
China for Children of All Ages

Demitasse

Demitasse sets have intrigued collectors for centuries. These dainty pieces, so reminiscent of our carefree childhood days, are extremely popular with Blue Ridge collectors. Prices have risen at an alarmingly rapid rate. Fortunately, additional sets with known patterns are now being recognized in this diminutive size, bringing greater numbers of sets on the market or into private collections than have previously been found.

Demitasse pieces have always been a favorite of collectors. As more Blue Ridge patterns appear, the quest intensifies for these delicate pieces.

RED TULIP, Demi Pot, Sugar, and Creamer, $220 set.

Above: Breakfast with SNIPPET.

Left: Breakfast with a set of YELLOW PLUME, Demi Cup and Saucer, $60. Cereal Bowl, $8 and Covered Toast, $120.

EMALEE, Demi Teapot, $145, Demi Cups & Saucers, $55 each set.

WILD MORNING GLORY, Individual Breakfast Set, $70. Colonial Faience backstamp.

MURIEL, Astor, Demi Teapot, Sugar, and Creamer, $255 set.

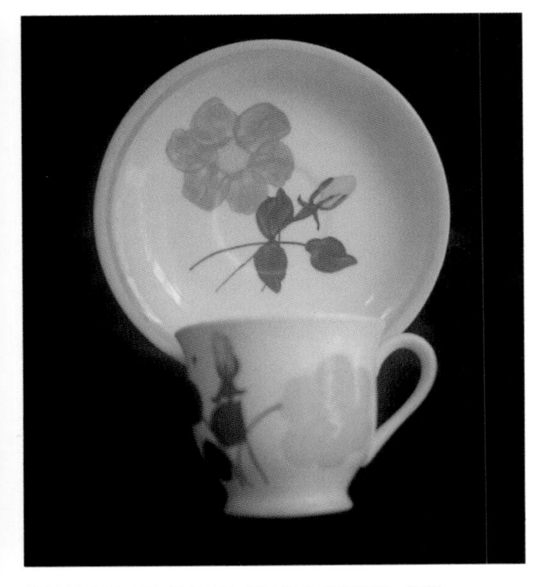

WRINKLED ROSE (YELLOW EDGE), Demi Cup and Saucer, $45.

SKEETER, Astor, Demi Cups and Saucers, $50 each set. Note the difference in hand painting.

WATERLILY, Demi Cup and Saucer, $50.

"BECK'S BLOSSOMS," Demitasse Tray, $75.

TEAL ROSEANNA, Demi Teapot, $125. Demi Cup and Saucer, $45.

SKITTER, Demi Cup and Saucer, $40, Demi Teapot, $120.

BAILEY BLUE, Demi Teapot, Sugar, and Creamer, $225 set.

TULIP BUDS, Breakfast Set, $500.

POINSETTIA, Demi Set, $300.

"PETIE," Demi Cup and Saucer, $50.

ANNADEL, Demi Cup and Saucer, $45.

BLUE HEAVEN, Demi Cups and Saucers, $55 a set.

Note MARDI GRAS Candlewick Demi Tray tucked in corner of this shelf.

DELTA DAISY, Demi Cup and Saucer, $30.

POM-POM VARIANT, Colonial, Demi Sugar, Demi Teapot, Cup, and Saucer, $275.

YELLOW NOCTURNE, Colonial, Demi Cup and Saucer, $35. Demi Teapot, $95.

YELLOW NOCTURNE, Demi Saucers, $15 each. Note the difference in painting.

VONDA, Demi Cups and Saucers, $40 each. Demi Teapot, $95. Look for a Big Cup and Saucer in this lovely pattern.

PETIT POINT, Child's set with original box, $1,000. Note "DEC. 4460," indicating the pattern number.

RICHARD, Demi Cup and Saucer, $45. CELANDINE, Demi Creamer, $45. CELENDINE, Demi Cup, $20.

PETUNIA, Demi Cup, $25. "RED PETUNIA," Saucer, $15.

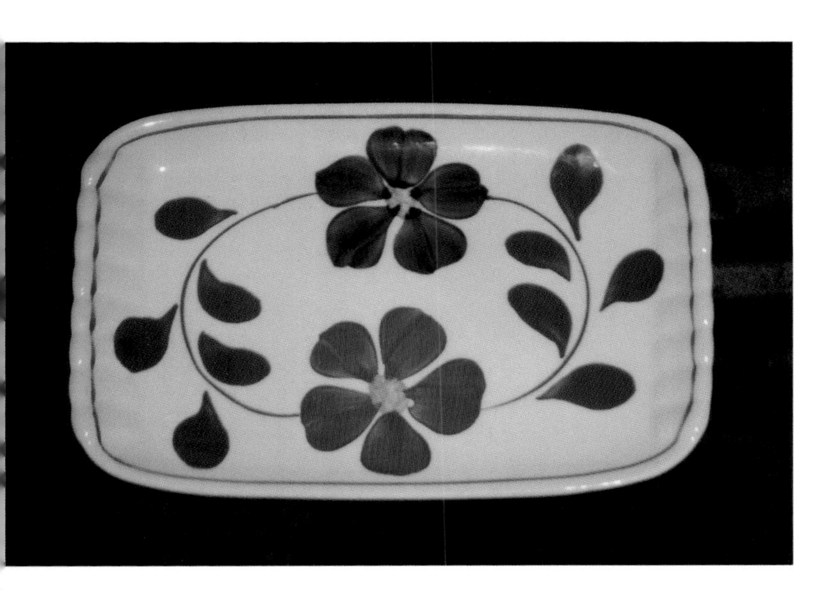

PETUNIA (LAUREL WREATH), Demi Tray, $125.

"COLETTE," Demi Tray, $100. This variation of PETUNIA was recently found with the name COLETTE as part of the backstamp. Slightly different leaves and coloring than PETUNIA.

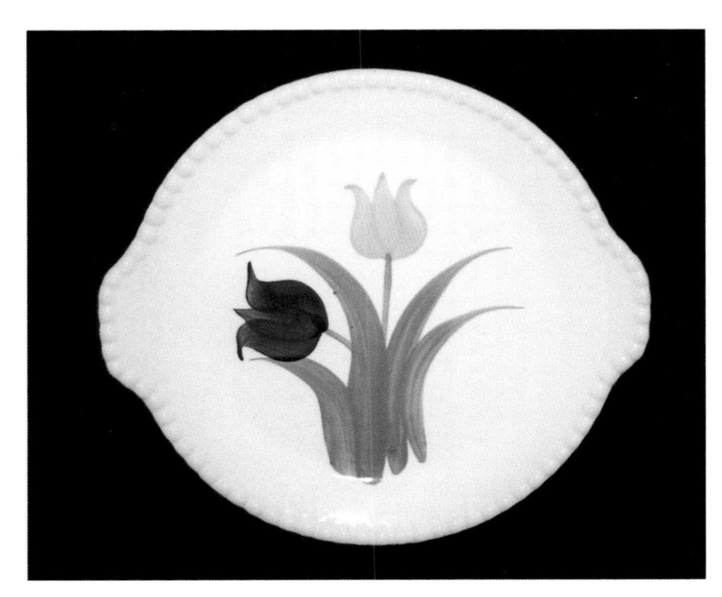

"JAN'S TULIPS," Candlewick, Demi Tray, $150.

ROMANCE, Demi Creamer, $55.

GOLDEN JUBILEE, Demi Cup and Saucer, $40.

ROSEMARIE, Demi Creamer, $55.

ROSEMARY, Demi Teapot, $125. Also called TULIP RING.

CHICKORY, Demi Cup and Saucer, $60.

GRANDMOTHER'S GARDEN, Demi Teapot, $140.

NOVE ROSE, Demi Cup and Saucer, $65.

Demis and Ovides are the new go-togethers.

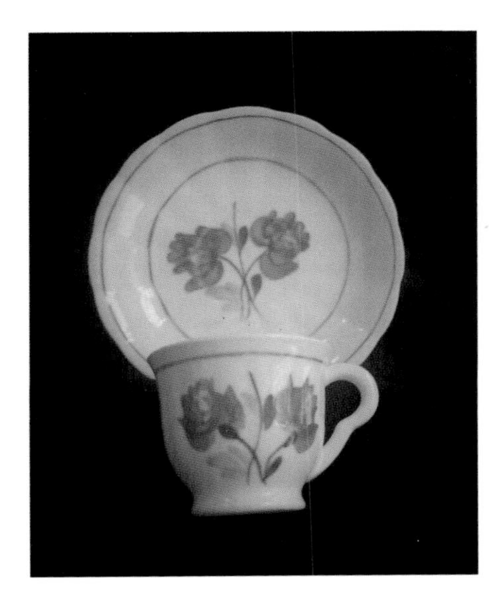

ROSY FUTURE, Colonial, Demi Cup and Saucer, $50.

BRECKENRIDGE, Colonial, Demi Sugar, $50.
FORGET-ME-NOT, Demi Sugar, $50.

COUNTRY GARDEN, Demi Cup and Saucer, $50.

BLACKBERRY LILY, Demi Cup & Saucer, $45.

BUGABOO, Demi Cup and Saucer, $45.

BACHELOR BUTTONS, Astor, Demi Cup and Saucer, $45.

MAGIC CARPET, Lace Edge, Demi Cup and Saucer, $60.

SYMPHONY, Demi Sugar, $50. DREAM FLOWER, Demi Creamer, $50.

PEONY, Demi Cup and Saucer, $50.

SPRING BOUQUET, Flared Sugar, $65. SPRING BOUQUET, Lace Edge, Demi Saucer, $40. This saucer has also been called SUMMERSWEET.

BLUEBELL BOUQUET (GREEN LEAF), Demi Teapot, $110.

"TICKLE," Demi Teapot, $125.

BLUEBELL BOUQUET (YELLOW LEAF), Astor, Demi Cup and Saucer, $45.

KAREN, Demi Creamer, Demi Sugar, Ovide Coffee Pot, Demi Cups and Saucers. An interesting combination.

PLUME, Astor, Demi Cup and Saucer, $50. Also found in yellow.

FANCY FREE, Demi Cup, $25. AMHURST, Demi Pot, $100.

BUTTERCUP (GYPSY),
Demi Cup, $20.

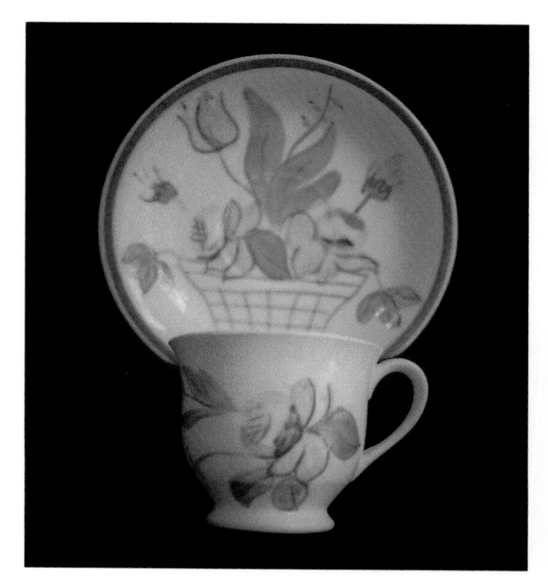

SONG SUNG BLUE, Clinchfield,
Demi Cup and Saucer, $60.

MISTRESS MARY, Astor, Demi Cup and Saucer, $125 set.

NOCTURNE (RED), Demi Cup, $25.

"BUDDIE'S DREAM," Astor,
Demi Cup, $35.

AMARYLIS, Demi Cup and Saucer, $55 set.

RED NOCTURNE VARIANT, Demi Tray, $125.

BEGGARWEED, Astor, Demi Saucer, $20. MOD TULIP, Colonial, Saucer, $20.

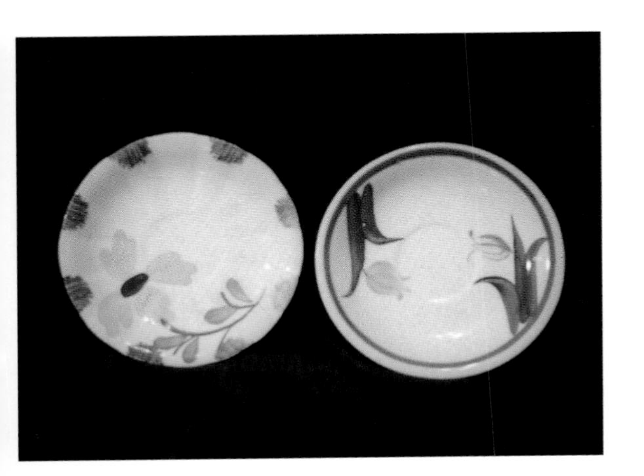

(Left to right) CYNTHIANA, Demi Saucer, $10. IRENE, Demi Saucer, $15.

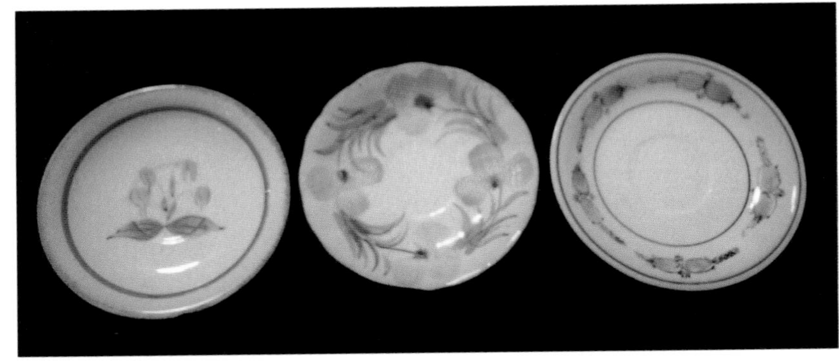

(Left to right) BLEEDING HEART, SWEET PEA, and BEADED CHAIN, Demi Saucers, $15 each.

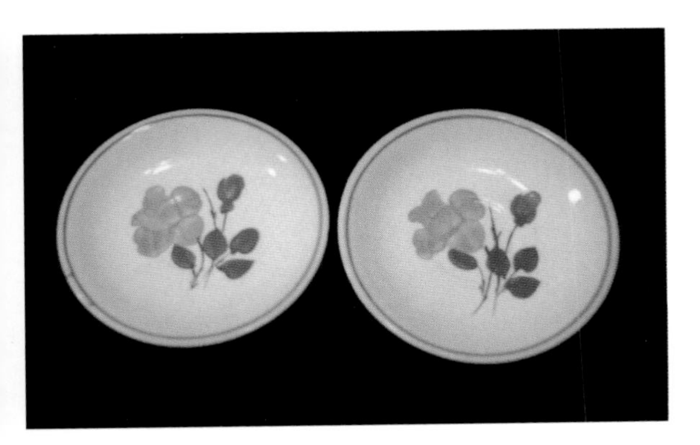

BRIDESMAID, Demi Saucers, $15 each. These can also be used with SUSANNA and VICTORIA.

"ANNELIESE," Clinchfield, Saucer, $25.

PRAIRIE ROSE, Pair of Demi Saucers, $20 each. Note variation in painting.

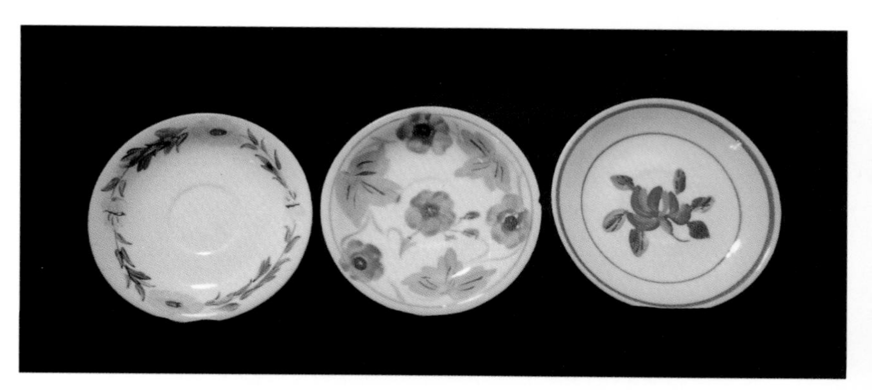

(Left to right)"BARBIE," IRISH MARY, HERB WREATH, Demi Saucers, $15 each

WILDWOOD FLOWER, Demi Cup, $25.

CHAMBLEE, Demi Cup, $25.

"JEAN'S DELIGHT," Colonial Demi Cup, $30, Saucer, $10, and Creamer, $25.

(Left to Right) REMEMBRANCE, Demi Saucer, $25, TAFFETA, Demi Saucer, $15.

STARFLOWER Demi Saucer, $15.

BEADED APPLE, Demi Sugar, $50. RED APPLE, Demi Saucer, $20.

LAVENDER FRUIT, Demi Cup and Saucer, $55.

MOUNTAIN CRAB, Demi Cup and Saucer, $50.

KAREN, Demi Saucer, $25, COREOPSIS, Demi Saucer, $15.

GRAPE SALAD, Demi Cup
and Saucer, $55.

Marshall Field & Company Shipping Box, Chicago, Illinois. A
child's tea set was shipped to a customer in Chicago.

FRENCH PEASANT, Demi Tray, $350.

BRITTANY, Demi Cup and Saucer, $125.

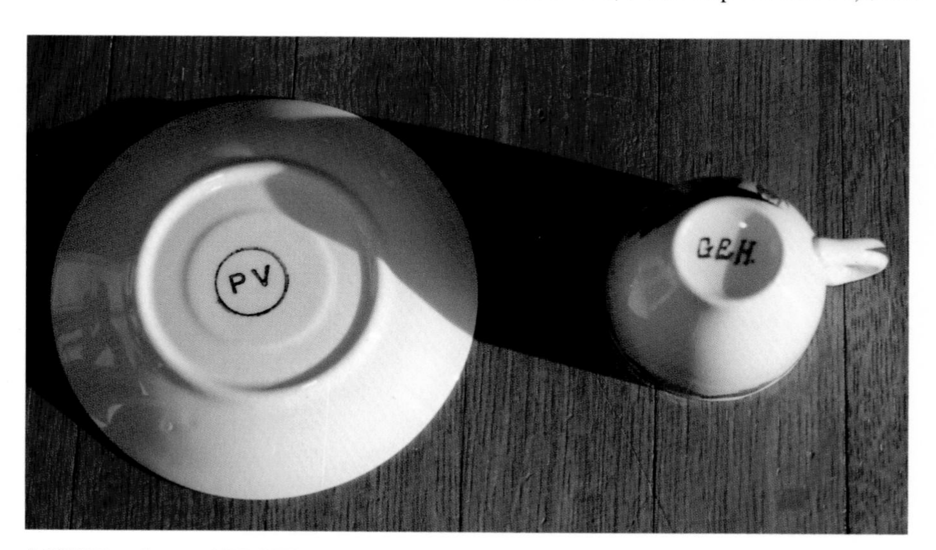

"GEH" on base of BRITTANY Demi Cup, Demi Saucer carries
"PV" backstamp. GEH believed to be initials of the artist.

FRENCH PEASANT, Demi Cups and Saucers, $125 each set.

Dinnerware

Dinnerware for children is truly for children of all ages. Today's prices for these pieces support this belief. Amazingly, many pieces have survived usage by children and storage in attics and basements. However, as with advertising pieces, collectors of children's dishes have to compete with collectors in the general collecting field.

YELLOW BUNNY, Astor, Plate, $350.

RED LEAF, Clinchfield, Clock, $25. JIGSAW, Feeding Dish, $350. FLOWER CHILDREN, Feeding Dish, $300. Children's pieces, although quite expensive now, were relatively inexpensive in the 1940s. They could be purchased through catalogs for $1 or less. They were also available coast to coast at stores like United China & Glass and D.E. Sanford.

YELLOW RABBIT, Skyline, Mug, $250. BUNNY HOP, Skyline, Mug, $250. Mugs are approximately 3" tall with small handles for small fingers to firmly grasp.

PINK PUPPY, Clinchfield, 7" Plate, $350. This plate with PIGGY BLUES Bowl and YELLOW RABBIT Mug were sold as a set (1940s Sears catalog) for less than $1.

PIGGY BLUES, 6" Bowl, YELLOW RABBIT, Mug, and PINK PUPPY, 7" Plate, Set $900.

MISS PIGGY, Astor, Plate, $350.

HUMPTY, Skyline, 8" Plate, $350.

PERFORMING SEAL, Skyline, Mug, $125. "HAPPY ELEPHANT," Skyline, Cereal Bowl, $300. Both pieces from three-piece Circus Set. Not pictured: "SILLY CLOWN" Plate, $300.

PERFORMING SEAL, Skyline, Mug, $125.

STACCATO, Mug, $100.

"ROBY," Piecrust, Plate, $500. Plate is marked #4212.

"Where is that Frog Plate?"

Ruth Cox Goodman and her cherished "ROBY" plate.

MISS MOUSE, Piecrust, Plate, $400.

Backstamp used for Miss Mouse.

PRACTICAL PIG, Mug can be seen on the top shelf (center).

"RUTH'S SET," Ruth received this set from a neighbor, who worked at the pottery. She remembers playing with them in her playhouse. Lucky little girl! "RUTH'S SET,"(Top to Bottom) "RUFFLES," LADY MOUSE, MISS MOUSE, AND "ROBY," $2,000.

FRUIT CHILDREN, Astor, Set of four, $1,400 (in original box.)

THREE LITTLE PIGS, Divided Feeding dish can be seen on top shelf on left.

FRUIT CHILDREN, Astor, Mug, $250.

FRUIT CHILDREN, Astor, Bowl, $300.

FRUIT CHILDREN, Astor, Plate, $300.

FRUIT CHILDREN, Astor, Feeding Dish, $300.

Chapter Ten
Novelties

Figurines

There is still debate as to the extent of Southern's production of figurines. Paper labels were in use at this time, so many pieces were left unmarked. Labels are removed and also wear away with time. Identification of Southern pieces is also complicated by the production of similar figurines by many area potteries.

Gold-trimmed Rabbit Figurine, unmarked.

Opening in back of rabbit allows for removal of cotton balls.

This little black rabbit looks like it could be Southern until...

You turn it over and look at the bottom.

"PINK RABBIT," Cotton Ball Keeper, unmarked.

"WHITE RABBIT," 4.5" tall, unmarked.

ELEPHANT FIGURINE, unmarked.

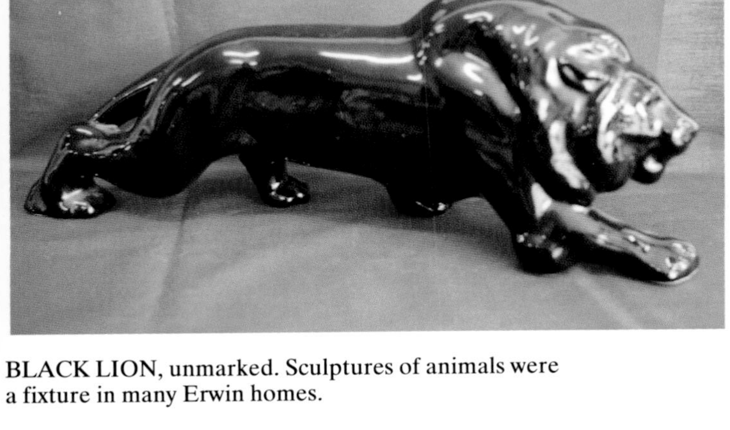

BLACK LION, unmarked. Sculptures of animals were a fixture in many Erwin homes.

ELEPHANT figurines vary in size.

RABBIT, Planter, unmarked.

ELEPHANT, Planter, unmarked.

Lunchtime Pieces

The flexibility of the management of Southern allowed employees to design and produce pieces for themselves. Many of these "lunchtime pieces" have survived and frequently surface. Many have been passed on to the next generation, sold privately or offered on the open market. It is nearly impossible to assign a value for or affix a price to these often one-of-a-kind pieces. We can, however, admire the talent that created them and enjoy their uniqueness.

SNAKE, Ashtray. A lunchtime creation.

"THE LAST SUPPER," Plaster Relief. Reportedly, only two of these were made. One was to be given as a Christmas present, and Floyd Jack Banner was allowed to keep the second one. The mold was destroyed.

"3 BUTTERCUPS & 3 FRIENDS," Ashtray, inscribed: "Banner B., Floyd F., and Jack J."

Back of ashtray, signed Ada Banner and dated April 19, 1943. Initialed V.S. in red.

"CREEKSIDE," Cake Plate, One of a kind.

Ashtray, signed Vergie Blevins, 1945, Jr. McCurry.

"NONE," Commemorative Jug with picture of Southern Potteries distributor.

Very large, hand painted "TURKEY FAMILY" Platter on Clinchfield shape. Unsigned and unmarked.

"SNOWY DAY," Skyline Plate, Signed "Wilma."

Chapter 11
The Tradition Continues

Although the city of Erwin continues to expand, the beautiful Blue Ridge Mountains are everlastingly unchanged.

Erwin Today

The Tennessee National Guard building has been the site of the Blue Ridge Pottery Show since 1997.

Outside the Main Street Mall. Yes, those really are Blue Ridge plates on the sign!

Sabre Brayton, owner of The Hanging Elephant, Gary Edmonds and Glenna Lewis, dealers in the mall.

(Left to right) Barbara Campbell, Billy Jack Campbell, and Maxie English, opened the Blue Ridge Pottery Store in March 1998.

Wanda Hashe, President of the Blue Ridge Pottery Club, and her mother, Dode, who worked at the pottery as a ware dresser.

Helen Linville was a painter for ten years. She recalls what each different brush was used to paint.

Dorothy Smith went to work at the pottery in 1922 at the age of sixteen.

Early Blue Ridge pieces on the mantel of the living room of the (Unicoi County) Heritage Museum.

China closet in newly redecorated Blue Ridge Room located in the (Unicoi County) Heritage Museum.

Table set with JOYCE in the museum. Table settings are changed to celebrate the seasons or a special day.

1996 Southern Pottery Reunion. Thanks, Employees of Southern Potteries!

Lisa Lingerfelt with her grandfather's Blue Ridge Pottery Club's Charter Plate. Richmond Barnett, 1980 Charter Member.

Sydney Nichols, a dealer at the Blue Ridge Show, with her extensive inventory of Blue Ridge.

Identification pins given to dealers at the Annual Blue Ridge Pottery Show. This show is held on the first weekend in October every year.

Betty Yates found the perfect plate during the Apple Festival.

Spin-Off Potteries

The Blue Ridge Mountains were alive with talented craftsmen and women, so a natural phenomenon resulted from the influence of the success of Southern Potteries. Several outstanding potteries produced memorable ware for us to admire and collect. Blue Ridge designs with individual embellishment are evident in the pieces produced at each pottery. These pieces have become collectible in their own right and have a following even outside the Blue Ridge Collector family.

Negatha Peterson's "MAMY" Cookie Jars, that portray the seasons or special days of the year, have found homes with many Blue Ridge collectors.

Erwin Pottery

TOPSY TURVEY, Cup and Saucer Wall Pocket.

"Negatha P." signature on most of the Erwin Pottery pieces. Sometimes, only the initials, "N.P." will be found.

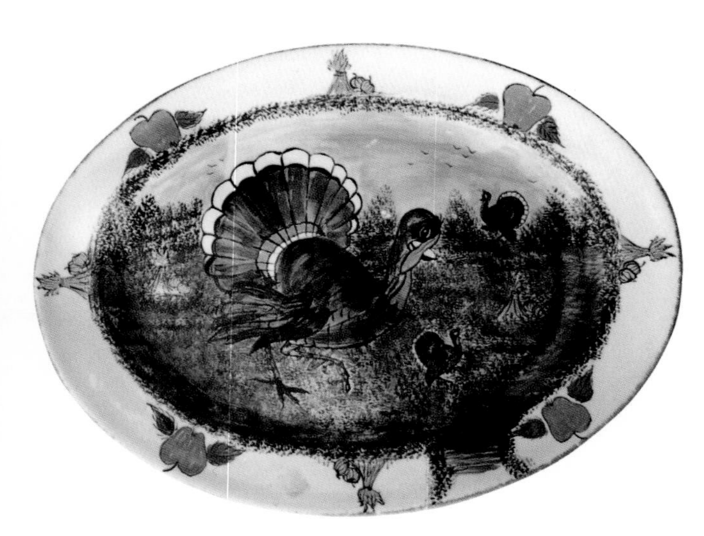

TURKEY Platter hand painted by Negatha Peterson of Erwin Pottery.

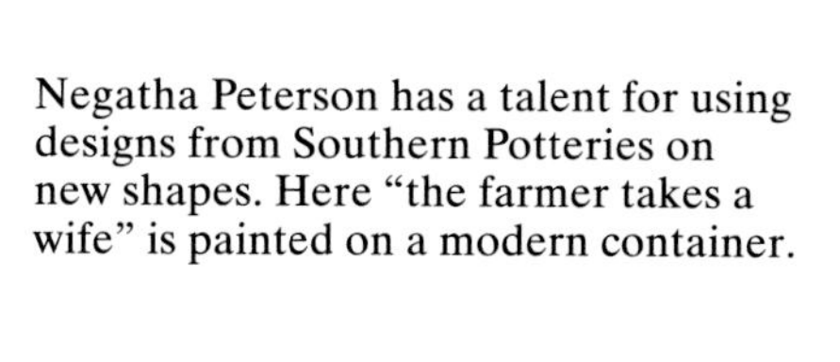

Negatha Peterson has a talent for using designs from Southern Potteries on new shapes. Here "the farmer takes a wife" is painted on a modern container.

Special "ROBY," Frog Plate designed by Negatha Peterson using the frog stamp.

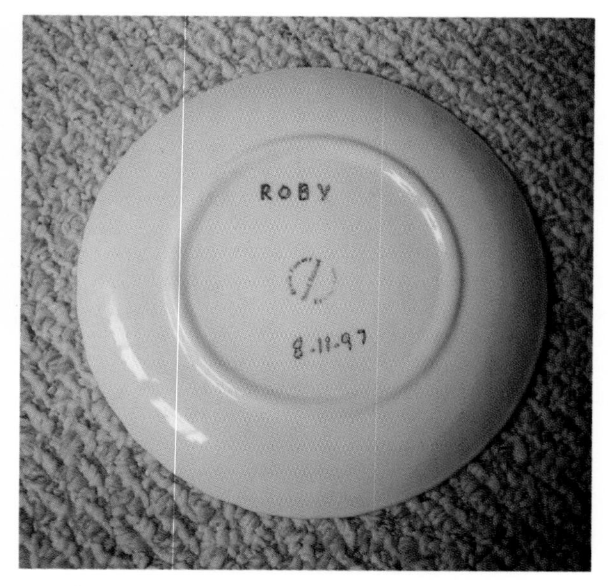

Back of the "ROBY" Frog Plate, on which is inscribed our grandson Roby's date of birth, August 11, 1997.

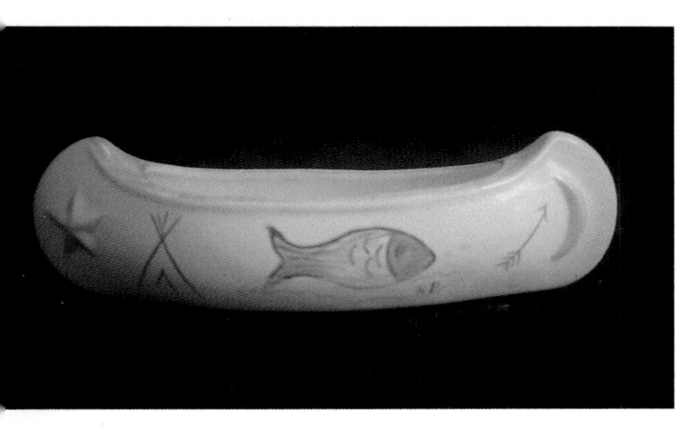

Erwin Pottery Canoe with TUNA SALAD motif.

Celebration Signs made by Negatha Peterson.

Cash Pottery

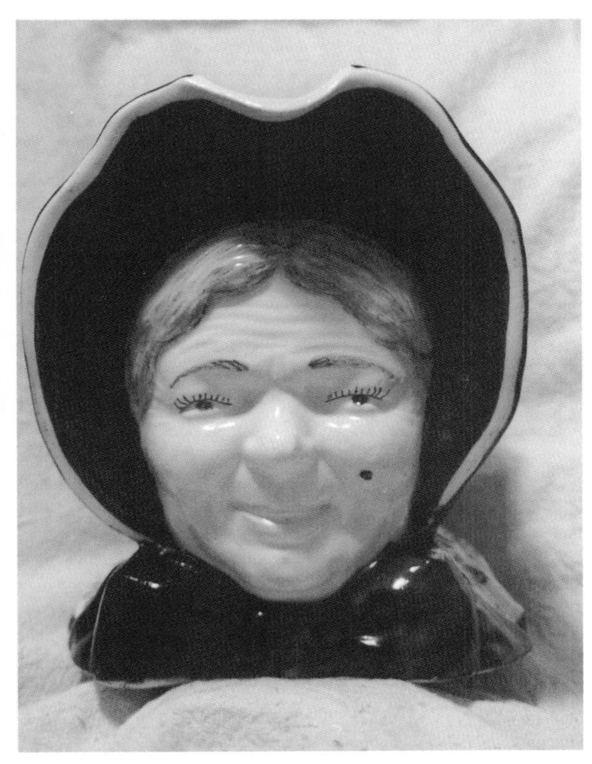

Cash version of the PIONEER WOMAN complete with beauty mole of the 1930s.

Bottom of Cash PIONEER WOMAN. Signed by Mildred Edwards 1945-1983, 6-22-83. Backstamp reads: Blue Ridge Pottery, Made by the Cash Family, Hand Painted, Erwin, Tenn.

An impressive collection of Cash Pottery items.

Beautiful DOGTOOTH VIOLET Fan Vase.

Duck on Nest by Cash. This piece comes apart to provide a lidded mixing bowl.

Clinchfield Artware Counter Sign.

Interesting Cash Carafe.

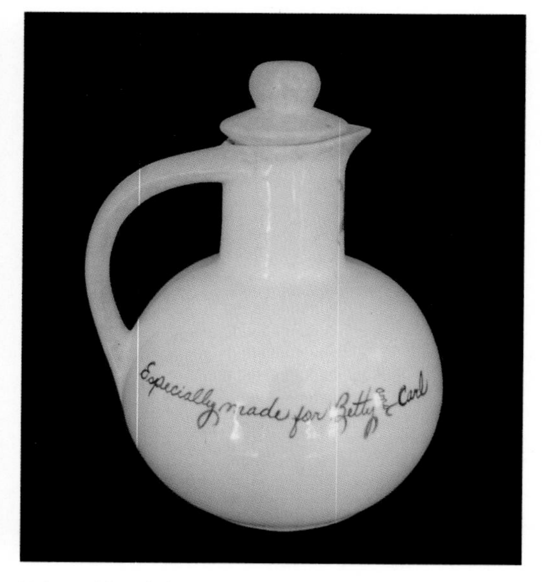

Other side of the Cash Carafe. Inscribed: "Especially made for Betty and Carl."

Corn Pitcher made by Cash.

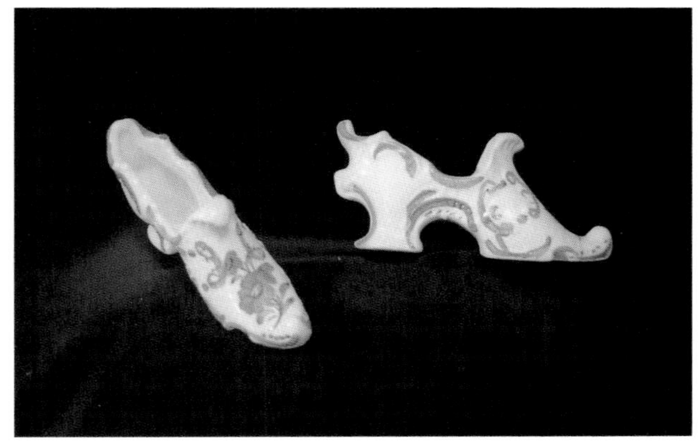

These dainty Slippers were made by the Cash family.

Well known Cash Pitcher and unusual short boot.

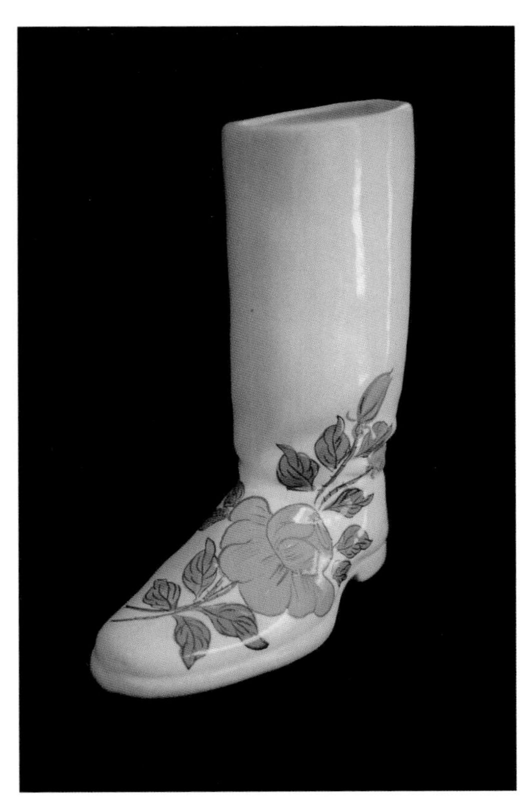

"YELLOW ROSE," Boot by Cash family. It is slightly larger than the Blue Ridge Boot Vases, measuring 8.25" in height with a foot span of 6.25".

Blue Cat Cup. There is a matching Yellow Cat Cup, which Mrs. Cash gave to a visitor for her children.

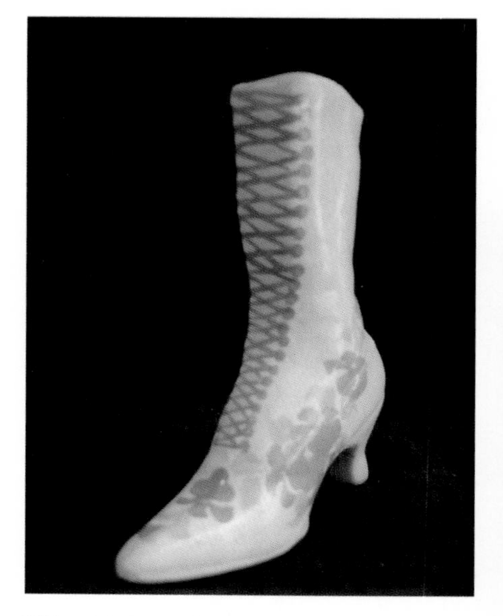

7" Lady's Laced Boot, stamped: Handpainted Erwin, Tenn.

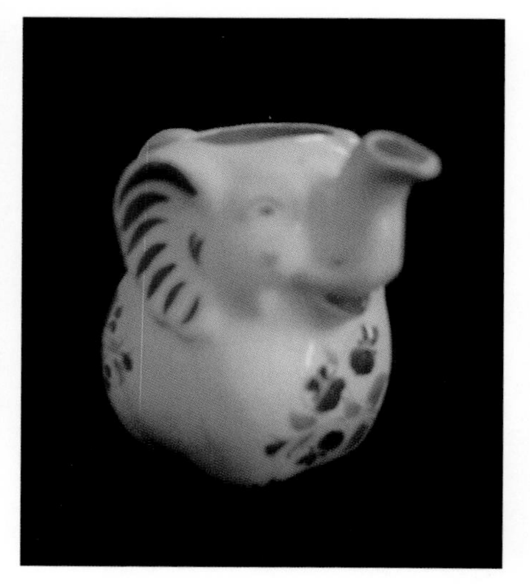

Elephant Creamer, stamped Clinchfield Ware Pottery, 1945.

Violet Pot, a very useful Cash piece. The top piece, which fits down into the bottom piece, has a hole in it to allow for the necessary drainage for violet plants.

Embossed Cow Milk Jug, one of the rarest Cash pieces.

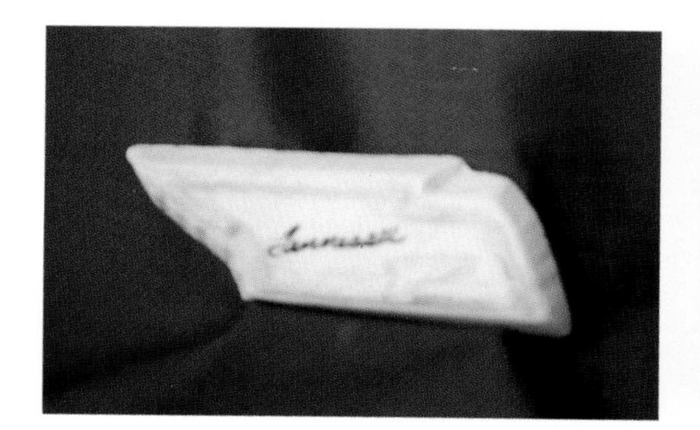

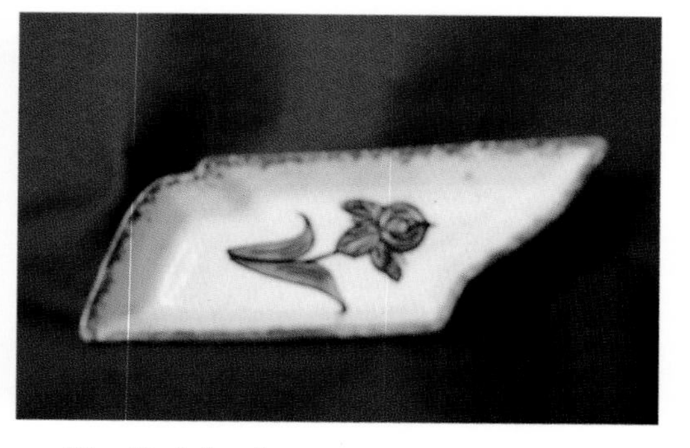

Souvenir Tennessee Ashtrays. The Cash family is well-known in the field of souvenir pottery.

Cash version of the popular BETSY. Eyelashes and rouge add a sweet touch.

This sweet little BETSY is unidentified but is presumed to be a Cash family piece.

"CHOCOLATE BETSY" made by Charles Duncan while he worked at the pottery. Note the shorter cap. Shown alongside a BLUE FLOWER BETSY by Blue Ridge.

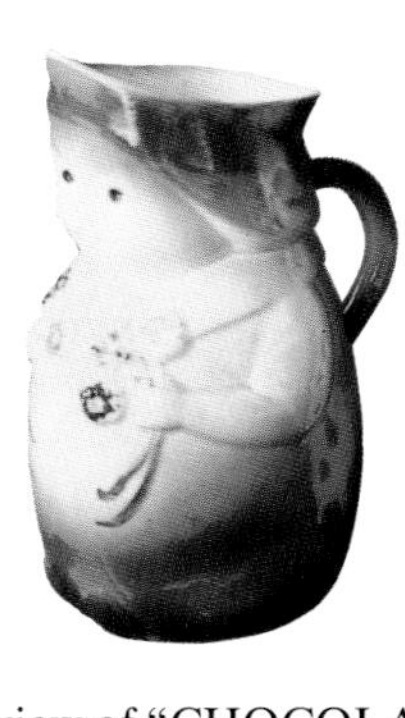

Side view of "CHOCOLATE BETSY." She is earthenware.

BETSY, painted by Mamie Cook.

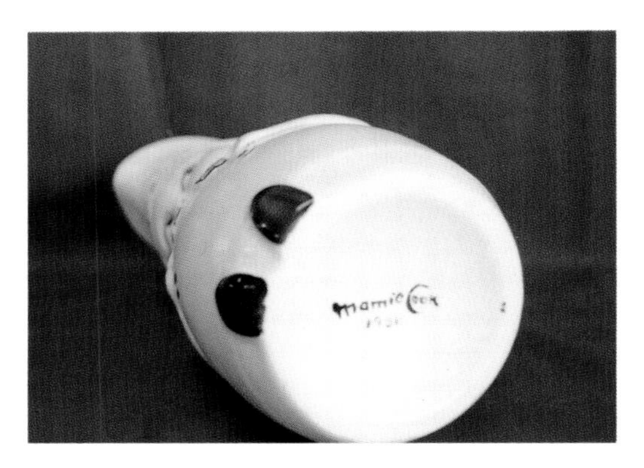

Bottom of BETSY, showing signature of Mamie Cook, 1931.

Two variations of TULIP BETSY. Note the eyelashes and full heart-shaped lips.

"DOT" BETSY with eyelashes and rouge. Origin unknown.

Creator of this smaller, daintier BETSY is unknown.

This cute little CHICK Pitcher with lots of blue eye shadow is unidentified.

Handled Vase, painted by Mamie Cook.

Bottom of Handled Vase, signed and dated M.Cook 1933.

Marie's

Heart-Shaped Box, painted by Marie.

4" Vase in dot-dash style by Marie. Backstamped: Marie's Handpainted Underglaze, Rockwood, Tenn.

Marie's backstamp.

Clouse Pottery

Trinket Boxes by Clouse.

Delicate dogwoods, gentle roses, and clinging vines are characteristics of Clouse pottery pieces favored by many collectors.

Clouse's BLEEDING HEART Demi Sugar and Creamer.

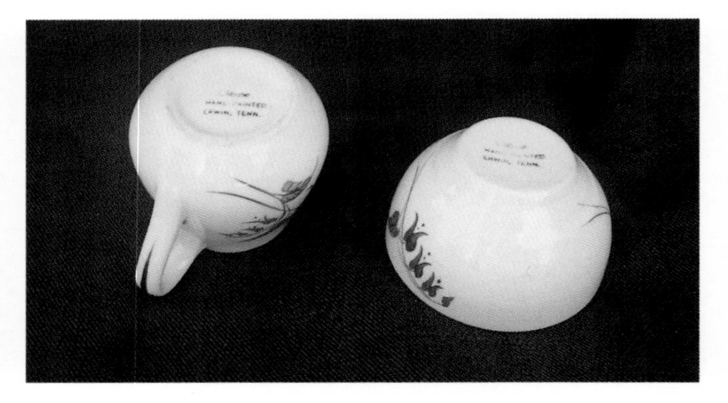

Backstamp on bottom of BLEEDING HEART pieces: Clouse, Handpainted, Erwin, Tenn.

Unmarked ORCHID Mini Jug is thought to be Clouse.

PANSY Demi Cup and Saucer by Clouse. Note Trellis-style handle. It's easy to understand why many Blue Ridge collectors of demitasse are eager to add Clouse sets to their collections.

Clouse PANSIES, Mini Vase.

CAMELLIA Pitcher and ROSE Boot, thought to be Clouse Pottery pieces.

PINK DOGWOOD, Small and Large Pitchers are popular items from Clouse Pottery. Note Trellis-type handle of the larger pitcher.

Unidentified

The "OPERA" china with the familiar PV backstamp was not made at Southern Potteries. Many different operas are depicted on the pieces in this series.

This adorable plate is often mistaken for Southern's PIECE OF CAKE pattern.

Small hanging ornaments were a favorite "take-home" item. Local potteries still make these.

Pieces with these Fondeville backstamps are not attributable to Southern Potteries.

Sculptured Fruit Pitchers are being made today by several potteries.

APPLE, Bell-shaped Salt Shaker.

Flower Pot with Blue Ridge-type flower. Reportedly, Southern planned to produce flower pots, but this story could not be confirmed. It would be nice to have these today to match our favorite Blue Ridge patterns.

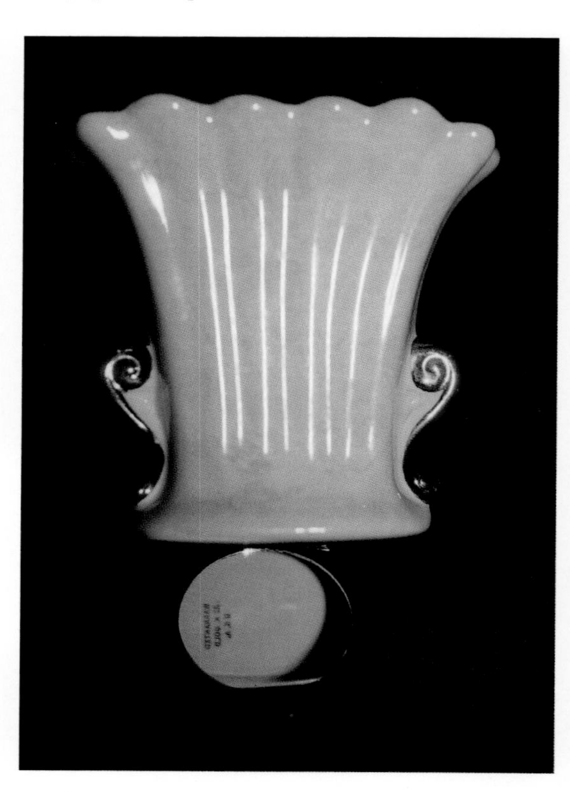

Pearl luster vase inscribed "Warranted 22K Gold USA."

Cruet Set in Blue Ridge-style pattern. The other side of the cruet set has been personalized for "The Howard's."

APPENDIX

Backstamps

Backstamps have been shown throughout the book in appropriate places. More common backstamps are repeated here for quick identification. Southern applied many backstamps to its pieces and often applied none. Trust your sixth sense when selecting pieces but always strive to be well-informed.

Script backstamp used for many years.

Backstamp found on some Clinchfield ashtrays.

Early backstamp, diamond shape.

Early backstamp in script.

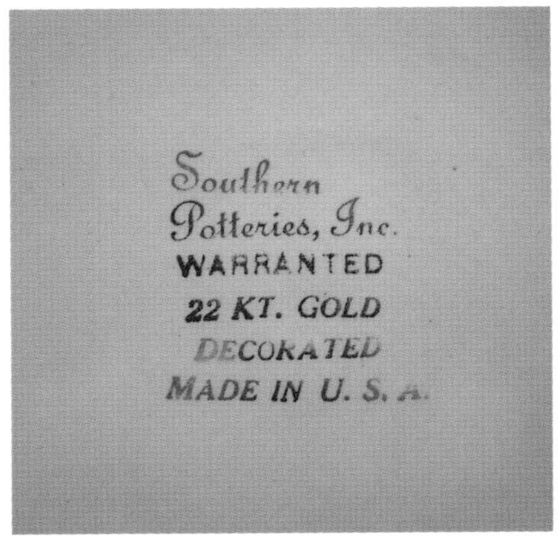

Backstamp often found on pieces decorated in gold.

Logo backstamp is most frequently found.

Later backstamp. Note Detergent Proof and Oven Safe

Later backstamp. Note Oven Proof and Fade Proof.

Label on back of TULIP TRIO plate showing number, 3216. These labels are helpful in identifying patterns.

Fondeville, New York backstamp.

King's backstamp (Jobber).

Berkshire backstamp is less common (Jobber).

UCAGCO (United China and Glass Company) backstamp.

Another UCAGCO backstamp (Jobber).

Bottom of SUNSHINE Bowl indicating pattern name.

WILD ROSE backstamp.

Early backstamp for Pattern # 2849-U.

Late backstamp for pattern #4561-3

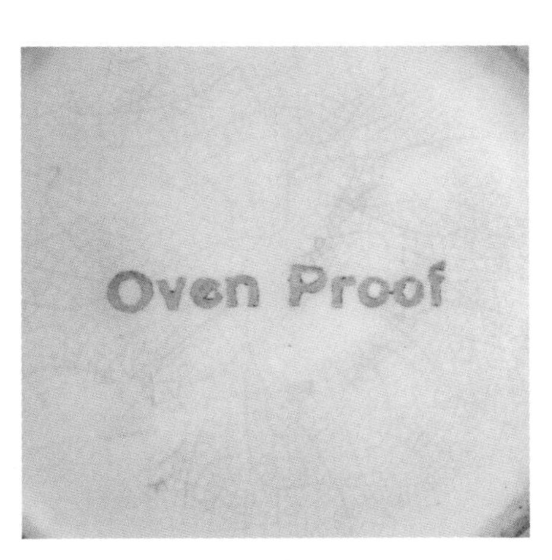

Cookware backstamp.

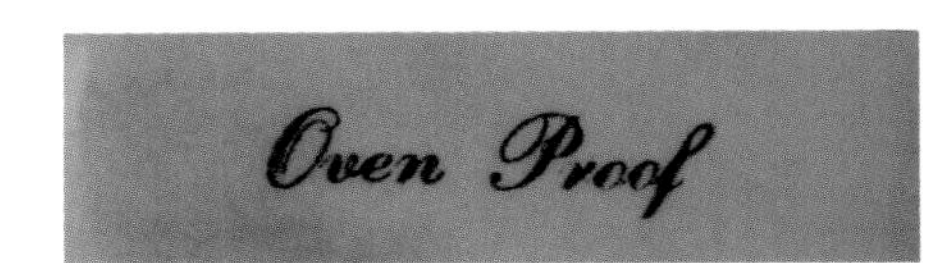

Cookware backstamp.

Many collectors of Blue Ridge like to acquire matching glassware. Advertisements in Sears and Ward's catalogs of the era afford us with some information in this area. Additional information on patterns for which glassware may be found is provided below.

Southern Potteries used a pattern numbering system to identify the many patterns they produced.

The Perry Antique Mall in Perry, Georgia carries a large inventory of Blue Ridge. Take a vacation from the Internet and shop the old-fashioned way!

BRAG (Blue Ridge Appreciation Group) Meeting. Warner Robins, Georgia. President: Marion Henry. Secretary: Sheila Ferguson.

Price Guide

Prices shown are based on collectors' valuations, current shelf prices, Internet and classified advertising prices for items in "mint" condition (no chips, cracks, crazing, glaze errors, paint smears or discoloration) at the date of publication. Prices may vary significantly based on condition, pattern, availability, popularity, and source. Most items pictured in this book have "average" prices shown in the captions for those pieces and represent neither the lowest nor highest prices for the items pictured.

PRICES SHOWN IN THIS BOOK ARE NOT INTENDED TO ESTABLISH ABSOLUTE VALUES ON ANY OR ALL ITEMS. THE AUTHORS AND PUBLISHER DISCLAIM ANY AND ALL LIABILITY FOR ANY LOSS RESULTING FROM RELIANCE ON PRICES HEREIN.

COMMON: Flora, Fruit, Lines, Plaids, Simple Designs
SPECIAL: Series such as Birds, Seasonal, Talisman, Nautical, Country
PROVINCIAL: French Peasant, Brittany, Normandy, Lyonnaise, Calais

DESCRIPTION	COMMON	SPECIAL	PROVINCIAL
PLATES:			
14" ROUND	$40-95		
12" ROUND (CAKE)	$30-45	$80-150	$150-200
10" ROUND (DINNER)	$15-40	$75-100	$100-125
9" ROUND (LUNCH)	$12-25	$65-90	$95-105
8" ROUND (SALAD)	$10-20	$60-80	$80-90
7" ROUND (DESSERT)	$10-20	$50-70	$70-80
6" ROUND (B&B)	$5-15	$40-60	$65-75
8" ROUND (SNACK)	$15-25	$50-75	$75-85
8" SQUARE	$25-40	$65-200	$100-150
7" SQUARE	$20-35	$60-200	$100-150
6" SQUARE	$20-35	$50-150	$90-120
PLATTERS:			
17"	$40-75	$100-350	$300-450
14"-15"	$25-65	$110-300	$275-350
12"-13"	$15-35	$100-250	$200-250
9-10"	$25-50		
CELERY	$15-50	$75-80	$90-100
CUPS & SAUCERS			
CUP	$5-15	$15-50	$40-70
SAUCER	$5-15	$15-50	$35-60
BIG CUP & SAUCER	$30-100	$125-150	$150-175
BOWLS:			
FRUIT-5"	$5-25	$25-35	$35-45
TAB -6"	$15-35	$35-45	$50-60
CEREAL-6"	$5-15	$25-40	$35-45
SOUP (7"-8")	$10-30	$35-60	$55-75
COVERED VEGETABLE	$60-95	$90-150	$110-175
VEGETABLE (ROUND 8"-9")	$15-50	$40-90	$75-100
VEGETABLE (DIVIDED)	$15-50	$40-60	$75-100
VEGETABLE (OVAL)	$15-45	$35-60	$65-85
SALAD BOWL (10"-13")	$55-85	$95-125	$125-150

DESCRIPTION	COMMON	SPECIAL	PROVINCIAL
SALAD FORK	$40-50	$100	$150
SALAD SPOON	$40-50	$100	$150
GRAVY BOAT	$15-35	$35-45	$75-85
CREAMERS:			
REGULAR	$15-35	$30-55	$50-85
PEDESTAL	$50-80	$85-95	$90-150
FLARED	$50-75	$75-85	$90-150
SUGARS:			
REGULAR	$15-35	$35-50	$55-90
OPEN	$10-30	$30-50	$50-75
PEDESTAL	$50-80	$70-90	$90-100
FLARED	$50-75	$75-85	$90-150
SALT & PEPPER SHAKERS (PAIR):			
BARREL	$20-35		
SKYLINE	$25-30	$45-60	$50-65
PALISADES	$25-30		
WOODCREST	$40-50	$50-65	
CHINA	$70-90	$90-125	$150-170
BLOSSOM TOP	$70-90		
BUD TOP	$70-90		
RANGE	$40-50		
CHICKEN		$110-175	
MALLARD		$300-450	
APPLE	$30-55		
PAINTED APPLE SHAPE		$25-30	
ACCESSORIES:			
BUTTER DISH (W/LID)	$30-45	$75-85	$100-150
BUTTER PAT/COASTER	$25-40	$50-95	$100-125
COVERED TOAST	$100-150	$150-180	$170-275
COVER ONLY	$50-70	$75-80	$100-125
PLATE ONLY	$50-70	$75-80	$100-125
CUSTARD CUP	$10-30	$40-75	$75-100

DESCRIPTION	COMMON	SPECIAL	PROVINCIAL
EGG CUP	$20-40	$50-60	$60-90
EGG PLATE		$75-125	
TILE/TRIVET	$45-60	$70-90	
CAKE LIFTER	$25-50		
SPOON REST	$30-60		

ACCESSORIES:

DESCRIPTION	COMMON	SPECIAL	PROVINCIAL
HEART RELISH	$50-80		
LEAF CELERY	$40-85	$75-100	$100-125
MAPLE LEAF RELISH	$60-110	$90-120	$120-145
LOOP HANDLE RELISH	$60-110	$90-120	
MOD LEAF RELISH	$60-110	$90-120	
CENTER HANDLE			
(4-SECT.)	$50-100	$90-120	
DEEP SHELL	$50-100	$90-120	$120-140
FLAT SHELL	$50-100	$90-120	$130-150
MARTHA SNACK TRAY	$100-150	$150-200	$225-275
LAZY SUSAN			
(W/HOLDER)	$600-1,000		
TIDBITS-1 TIER	$20-30	$40-50	$80-100
TIDBITS-2 TIER	$25-35	$50-80	$100-125
TIDBITS-3 TIER	$35-45	$60-90	$120-140
SERVERS			
(METAL/WICKER)	$35-45		
CLOCKS	$25-45	$50-75	$75-100

TEAPOTS:

DESCRIPTION	COMMON	SPECIAL	PROVINCIAL
COLONIAL	$90-250	$200-350	$250-450
PIECRUST	$70-90		
SKYLINE	$70-95	$200-250	$300
PALISADES	$85-120		
WOODCREST	$130-160	$200-250	
ROPE HANDLE	$80-120		
BALL	$110-150	$200-250	$300
MINI-BALL	$150-175		
SQUARE ROUND	$125-140		
CHEVRON	$130-150		
FINE PANEL	$125-140		
SNUB NOSE	$100-150		
CARAFE	$35-50		
CARAFE W/STAND	$65-75		
COFFEE POT (OVIDE)	$95-150	$150-300	$300-350
CHOCOLATE POT	$200-225	$275-350	$350-450
CHOCOLATE TRAY	$500-700	$750-900	$900-1,000

DEMITASSE:

DESCRIPTION	COMMON	SPECIAL	PROVINCIAL
TEAPOT	$75-175	$150-250	$225-275
SUGAR	$25-60	$50-80	$80-100
CREAMER	$25-60	$50-80	$80-100
CUP	$20-45	$25-65	$40-75
SAUCER	$15-30	$20-55	$35-65
TRAY	$100-135	$150-200	$200-275

PITCHERS:

DESCRIPTION	COMMON	SPECIAL	PROVINCIAL
ABBY	$75-100		
ALICE	$115-175		$260-290
ANTIQUE 3.5"	$110-135		$175-275
ANTIQUE 5"	$60-100		$160-170
BETSY			
(EARTHEN WARE)	$85-135		
BETSY (GOLD)	$325		
BETSY (CHINA)	$150-275		
CHICK (CHINA)	$115-145		
CHICK			
(EARTHEN WARE)	$50-75		
CLARA	$75-100		$165-180
GRACE	$75-125		$165-180
HELEN	$80-175		
JANE	$100-150		$175-190
MARTHA	$40-60		
MILADY	$135-200		$300-325
PALISADES	$40-90		
REBECCA	$120-200		
SALLY	$100-200		
SCULPTURED FRUIT	$70-95		
SPIRAL 4.5"	$100-150		
SPIRAL 7"	$75-150		
VIRGINIA 4.5"	$100-150		
VIRGINIA 6.5"	$75-125		$165-180
WATAUGA	$200-300		

VASES:

DESCRIPTION	COMMON	SPECIAL	PROVINCIAL
BOOT	$75-125		
BOOT (GOLD)	$125-150		
BUD	$175-300		
BULBOUS	$65-100		
HANDLED	$100-200		
RUFFLE TOP	$75-100		$175-190
TAPERED	$100-130		$150-160

LAMPS:

DESCRIPTION	COMMON	SPECIAL	PROVINCIAL
HANDLED	$150-200		
GRECIAN	$200-300		
TEAPOT / COFFEE POT	$100-150		
WALL	$75-125		

BOXES (W/LIDS):

DESCRIPTION	COMMON	SPECIAL	PROVINCIAL
CANDY		$150-275	
CIGARETTE			
(FLORAL/FRUIT)		$50-100	
CIGARETTE			
(PROVINCIAL)		$125-200	
CIGARETTE			
(SPECIAL)		$75-120	
ASHTRAYS			
(INDIVIDUAL)		$20-50	
DANCING NUDE			
(MARKED)		$300-400	
MALLARD		$500-750	
ROSE STEP		$200-250	
SEASIDE		$175-225	
SHERMAN LILY		$400-500	
VANITY (POWDER)		$120-200	

DESCRIPTION	COMMON	SPECIAL	PROVINCIAL

CHILDREN'S DINNERWARE:

DESCRIPTION	COMMON	SPECIAL	PROVINCIAL
BOWL		$150-250	
DIVIDED FEEDING BOWL		$200-300	
FEEDING DISH		$150-300	
MUG		$150-250	
PLATE		$150-300	

COOKWARE:

DESCRIPTION	COMMON	SPECIAL	PROVINCIAL
RAMEKIN, SMALL (W/LID)		$35-45	
RAMEKIN, LARGE (W/LID)		$55-70	
COVERED CASSEROLE		$40-60	
BAKING DISH		$35-50	
BATTER PITCHER		$75-125	
SYRUP PITCHER		$50-100	
MIXING BOWLS (SET OF FIVE)		$100-175	
MIXING BOWLS (SET OF 3)		$75-125	
PIE BAKER (ROUND)		$35-50	
LEFTOVER CONTAINERS (SET OF 3)		$75-150	

CHARACTER JUGS:

DESCRIPTION	COMMON	SPECIAL	PROVINCIAL
DANIEL BOONE		$600-850	
PAUL REVERE		$600-850	
PIONEER WOMAN		$600-850	
INDIAN			$900-1,100
PIONEER WOMAN (MINIATURE)			$3,500

ARTIST SIGNED:

DESCRIPTION	COMMON	SPECIAL	PROVINCIAL
PLATTERS		$1,000-2,500	
PLATES		$600-1,500	

CHARM HOUSE:

DESCRIPTION	COMMON	SPECIAL	PROVINCIAL
CREAMER		$90-125	
PITCHER		$175-225	
RAMEKIN (SMALL)		$100-150	
SALT & PEPPER (PAIR)		$125-150	
SUGAR		$90-125	
TEAPOT		$150-250	

GOODHOUSEKEEPING:

DESCRIPTION	COMMON	SPECIAL	PROVINCIAL
CREAMER		$40-60	
SALT & PEPPER (PAIR)		$100-130	
SUGAR		$40-60	
TEAPOT		$150-225	

ADVERTISING:

DESCRIPTION	COMMON	SPECIAL	PROVINCIAL
ASHTRAYS (DECAL ONLY WITH BACKSTAMP)			$50-60
ASHTRAYS (DECAL WITH PAINTED TRIM)			$60-95
BREAD & BUTTER		$25-50	
BLUE RIDGE (LOGO PLATE)			$300-600
BOWL (GRANNY)		$35-45	
COUNTER SIGN (BLUE RIDGE)			$200-250
DINNER PLATE		$30-60	
GEORGE WASHINGTON (DECAL PLATE)			$200-300
PITCHERS		$60-125	
PRIMROSE CHINA (LOGO PLATE)		$300-350	
ROBERT E. LEE (DECAL BOWL)			$200-300
ROBERT E. LEE (DECAL PLATE)			$200-300
TALISMAN WALLPAPER (LOGO PLATE)			$300-400

CLINCHFIELD (DINNERWARE):

DESCRIPTION	COMMON	SPECIAL	PROVINCIAL
PLATTER		$20-50	
PLATE (DINNER)		$10-50	
PLATE (B & B)		$5-10	
CUP & SAUCER		$10-25	
SUGAR		$15-25	
CREAMER		$15-25	
VEGETABLE BOWL		$45-55	
PITCHER		$40-95	
SALT & PEPPER (PAIR)		$45-75	

Pattern Numbers

107 CLOWNS (TEMP#)
343TEM "LAZY DAY FARMS"
350-1 MALLARD BOX
593TEM ATHENS
717A "OUCH"
730U POND IRIS
1213US "KRIS LEIGH"
1542UC GYPSY FLOWER
2387 MARDI GRAS & MARDI GRAS VARIANT
2455 MOUNTAIN ROSE
2639U CROSSBAR
2662 SUNGOLD #2
2667 "ALLISON"
2673 MARY
2754 LAURA
2756 CHRYSANTHEMUM
2788 BLUEBELL BOUQUET
2790 ROCK ROSE & VALLEY BLOSSOM
2823 YELLOW NOCTURNE & SUNFLOWER
2934U POLKA DOT
2947 GARDEN LANE & JUNE BOUQUET
3007 FLOWERING BERRY
3043 COUNTY FAIR (CHERRY)
3078 WRINKLED ROSE
3090 CHINTZ
3093 SEREPTA
3123 #7 LAZY DAZY
3149 CUMBERLAND
3196 DOGWOOD & APPLE TART
3238 FRENCH PEASANT
3245 FRENCH PEASANT
3254 MARINER & SAILBOAT
3261 WATERLILY
3272 MOUNTAIN ROSE
3274 COUNTRY ROAD
3276 GRANDMOTHER'S GARDEN
3280 AROUND ROSEY
3329 CHERRY DROPS
3305 COUNTRY ROAD VARIANT
3324 WILD IRISH ROSE
3334 FALLING LEAVES
3340 PERIWINKLE
3343 POINSETTIA
3353 SWEET CLOVER
3361 HOPSCOTCH
3377 ROSE OF SHARON
3387 GYPSY (BUTTERCUP)
3389 MICKEY
3398 GARDEN GREEN & POMONA
3423S APPLE TRIO
3424 SPIDERWORT
3451 MAUDE
3458 YELLOW PETUNIA
3461-4 FRUIT FANTASY
3521 FLORAL MEDLEY
3534 RED HILL
3535 SUNBRIGHT
3542 LEXINGTON
3545 POINSETTIA
3546 COUNTRY GARDEN
3547 RUGOSA
3567 VIOLETS (FOR JEAN WK)
3558 WILD STRAWBERRY
3583 HAT DANCER
3584 SENORITA
3585 COCKFIGHT
3586 CHICKEN MAN
3587 MEXICAN
3588 PEANUT VENDER
3589 MEXICAN
3590 MEXICAN WOMAN
3599 HOLLYBERRY & HOLLY
3623 "THE PRIZE"
3637 BEADED APPLE
3639 ZINNIA
3640 SWEET PEA
3642 KATE
3643 TAFFETA
3648 CHICKORY
3661 BOURBON ROSE
3663 PEONY BOUQUET
3668 JUNE BOUQUET
3671 ROSE HILL
3687 "NEGATHA" & SUNDAY BEST
3688 WILD ROSE
3689 "SCARLET"
3694 "OUR LISA"
3715 BECKY
3721 CHERRY BOUNCE
3733 MIRROR MIRROR & TRIPLE TREAT
3735 AUTUMN APPLE
3736 ORCHARD GLORY
3738 MIRROR IMAGE
3739 PAPER ROSES

3743 FRUIT PUNCH
3747 SAVANNAH
3755 FULL BLOOM
3760 FLOWER RING
3761 CAROL'S ROSES
3768 RUTH ANNA
3769 TIGER LILY
3770 CHERRY BOUNCE
3773 CRAB APPLE
3775 KIBBLER'S ROSE
3777 ROSETTE
3780 BIG APPLE
3782 CHERRY COKE
3783 EASTER PARADE
3787 SUNFIRE
3810 SUNFIRE
3826 HOLLYHOCK
3837 FREEDOM RING
3850 MOUNTAIN IVY
3852 RIDGE DAISY
3855 SWISS DANCERS
3862 ANGELINA
3863 DAYDREAM
3866 PANDORA
3871 DRESDEN DOLL
3872 ROXALANA
3875 FOLK ART FRUIT
3885 AMELIA
3886 DUTCH BOUQUET
3887 DELICIOUS
3888 AMELIA
3889 ECHOTA & TWIN FLOWER
3892 FLOWER CHILDREN PLATE
3896 MOD TULIP
3897 SHOO FLY
3898 MOUNTAIN BELLS
3901 DEWBERRY
3908 BOUTONNIERE
3910 SUNFIRE TEAL
3913 DIXIE HARVEST
3916 FREEDOM
3919 SOUTHERN CAMELLIA
3920 LIGHTHEARTED
3924 NOCTURNE (3 BUDS)
3925 DAHLIA
3937 SYMPHONY
3951 POPPY DUET
3951 SUNGOLD#1
3953 POPPY DUET
3954 POM POM
3963 JOANNA
3966 CHERRY WINE
3968 HIGHLAND IVY
3969 DELLA ROBBIA
3985 ROSE RED
3986 TRICOLOR VARIANT
3988 CARNIVAL
3998 FUSCHIA
4001 PINK DAISY
4002 RUTLEDGE
4012 SUNNY
4013 WINDFLOWER
4014 BETTY
4022 "TWIN APPLES"
4023 RUTLEDGE
4024 MOUNTAIN MEADOWS/ALICIA
4030 BACKYARD BLEEDING HEARTS
4032 "VIVIAN"
4042 GREENBRIAR
4043 SUN BOUQUET
4047 WILLOW
4049 WHIRLIGIG
4050 EVENING FLOWER
4051 FANTASIA
4052 SUNNY SPRAY
4057 FOREST FRUIT
4073 RIDGE HARVEST
4074 PETAL POINT
4076 FIRST LOVE
4077 FRAGERIA
4082 WINNIE
4084 SUN BOUQUET
4085 CHANTICLEER
4089 ALLEGRO
4093 GRAY SMOKE
4094 JAPANESE WALLFLOWER
4096 GREEN EYES
4113 SPRING BEAUTY
4116 FOXGRAPE
4119 SUSAN
4120 DOE EYES
4128 BITTERSWEET
4129 VIBRANT/ALLEGRO
4141 VIOLET SPRAY
4145 FLIRT
4146 RUSTIC PLAID

4149 MAROON PLAID
4153 LORETTA
4158 PLANTATION IVY
4160 BETHANY BERRY
4163 MIDAS TOUCH
4174 "NORTHERN DOGWOOD"
4175 SOUTHERN DOGWOOD
4176 FLAPPER
4178 "TWISTED RIBBONS"
4184 MOD OAK LEAF
4192 BALTIC IVY
4193 QUAKER APPLE
4195 YELLOW DOGWOOD
4201 BLOSSOM TREE
4206 CHICKEN PICKINS
4212 "ROBY"
4225 MAYFLOWER, MAYFLOWER W/BLUE, & CHARLOTTE
4226 CHERRIO
4242 GRANNY SMITH APPLE
4243 SCATTER PLAID
4245 SABILLA
4250 FRENCH VIOLETS & VIOLET SPRAY
4261 DUCK IN HAT
4262 BOUNTIFUL
4266 FLOWER FANTASY
4274 NORMANDY
4275 CAROLINE
4276 APPLE JACK
4277 WEATHERVANE
4288 BAMBOO W/GREEN (SARASOTA)
4289 BAMBOO
4296 SPRAY
4298 OAKDALE
4299 MARK
4309 APPLE CRUNCH
4317 RED BARN
4321 TROPICAL W/FLOWERS
4324 RED RING
4325 STANHOME IVY
4327 LOVELY LINDA
4331 ORCHID
4332 GOLDEN SAND (GOLD PATRICIA)
4333 NIGHT FLOWER
4335 GINGHAM APPLE
4336-X CALICO FARM (APPLECHECk FARM)
4338 FAIRMONT
4340 APPLE JACK
4340 CAROLINE
4341 ORCHID
4344 TROPICAL
4356 KIBBLER'S ROSE
4356SL DESERT FLOWER
4361 "SPRAY AROUND"
4365 SUNNY SPRAY & SERENADE
4370 CHRISTMAS DOORWAY
4371 DAFFODIL
4376 WIND MILL
4378 SUNRISE (SKYLINE)
4381 PETIT POINT
4382 JUNE APPLE
4386 EVENTIDE & FARMHOUSE
4387 MING TREE (GREEN)
4390 QUILTED FRUIT
4391 SUNRISE (WOODCREST)
4407 MING TREE (WHITE)
4409 SUNNY–UP
4413 SPRING HILL TULIP
4419 TIC TACK
4422 RED ROOSTER
4423 STRAWBERRY SUNDAE
4432 "SIMPLY WILD CHERRY"(#1), & "EARTHY WILD CHERRY" (#3)
4436 CHRISTMAS TREE
4445 HARVESTTIME
4456 PILGRIMS
4460 PETAL POINT & PETIT POINT
4479 MOUNTAIN SWEETBRIAR
4486 MOSS ROSE
4489-S LUNA
4497 GOOD LUCK-RAGTIME
4499 ROUNDELAY
4501 ROUNDELAY-GREEN & BROWN
4512 APRIL
4523 BARDSTOWN
4527 COUNTRY FRUIT
4532-Y THISTLE
4534 GINGHAM FRUIT
4536 COUNTRY FRUIT
4553 CLAIRBORNE
4560-Y CLOVER
4561S "TWIRLING TRIO"
4566 STRAWBERRY RING
4572-Z SODDY DAISY
4590 TURKEY WITH ACORNS
4591 CHEROKEE ROSE

4591S CHEROKEE ROSE
4614 SILHOUETTE
4616 NESTING BIRDS
4617 CATTAILS
4617A CATTAILS
4621 SHADOW FRUIT
4623-P RUTLEDGE
4633F BRIAN
4634 ANDANTE
4639 SPIDER WEB PINK
4669 WILD CHERRY #2
4616 NESTING BIRDS
4659 JENNIFER
4681 MAYFLOWER
4699 "AQUA WILD CHERRY" (#4)
4700 PINKIE
4703 RIDGE HARVEST
4716 KIMBERLEY
4718 GRASS FLOWER
4733 WHIPSTITCH
4742 COCK A DOODLE
4753 WILLA
4754 GOLDEN BELLS
4756 HUGH KIBLER
4853 STARFLOWER
4853 STACCOTO
5001 RED NOCTURNE
5005 PEGGY
5006 NOLA & FROLIC
5015 ENDEARING
6001 "BUTTERFLY"
6002 "OASIS"
7051 MARINER
7443 "DAY LILY"
4664 STRAWBERRY SUNDAE
4670 UNAKA

Matching Glassware for Blue Ridge

The following Blue Ridge patterns had matching glassware (11 oz. Tumbler, 6 oz. Juice, and 7.5 oz. dessert cups):

Bali Hai (or Bamboo)*
Calico (or Country Fruit)
Chintz
Country Charm (Red Barn)
County Fair
Crab Apple**
Cumberland
Dahlia*
Damascus
Evening Flower****
Garden Lane***
Green Briar
Mountain Cherries*
Mountain Ivy
Petal Point
Red Rooster*
Ridge Daisy**
Ridge Harvest
Ridge Ivy (or Baltic Ivy)
Sun Bouquet
Susan
Vibrant (found in Federal)
Whirligig*

Note: A Libby pitcher (with a rolled base) that matches Ridge Daisy has been found.

* Photos of these items reflect only 11 oz. Tumbler.
** Also Federal flat water (12 oz., 5.25' high) and juice (5 oz., 3 5/8"high).
*** Only seen in Federal flat water (12 oz., 5.25" high) and juice (5 oz., 3 5/8" high).
**** See *Blue Ridge China Today*(Ruffin), page 117.

Resources

Blue Ridge Pottery Club
208 Harris Street
Erwin, TN 37650
(Wanda Hashe)

Blue Ridge Quarterly
203 St. Raymond Place
Kathleen, GA 31047-2125
(Frances & John Ruffin)
(jruff@peachnet.campus.mci.net)

National Blue Ridge Collector's Directory
3737 Tyler St.
Columbia Heights, MN 55421
(Jay Parker)

National Blue Ridge Newsletter
144 Highland Dr.
Blountsville, TN 37617
(Norma Lilly)

References

Books and Journals

Audubon, John James. *Birds of America*, New York: Macmillan Publishing, 1953.

Cunningham, Jo. *Collector's Encyclopedia of American Dinnerware*. Paducah, Kentucky: Collector Books, 1992.

Keillor, Winnie. *Dishes What Else?* Frankfort, Michigan, 1983.

___________. *Dishes What Else? (Supplement)* Frankfort, Michigan, Undated.

Kovel, Ralph and Terry. *Kovel's New Dictionary of Marks*. New York: Crown Publishers, 1986.

Larkin, Diane. "Southern Potteries, Erwin, Tennessee's Dinnerware Legacy," *Blue Ridge Country*, May/June 1994, pp. 18-19, 51.

Nelson, Maxine Feek. *Collectible Vernon Kilns*. Paducah, Kentucky: Collector Books, 1994.

Newbound, Betty and Bill. *Blue Ridge Dinnerware*. Puducah, Kentucky: Collector Books, 1984, 1989.

_________. *Collector's Encyclopedia of Blue Ridge Dinnerware*. Paducah, Kentucky: Collector Books, 1994, 1998.

Robbins, Brunn, and Zim. *Birds of North America*. New York: Golden Press, 1966.

Ruffin, Frances and John. *Blue Ridge China Today*. Atglen, Pennsylvania: Schiffer Publishing, 1997.

Veiller, Lawrence. "Industrial Housing Developments in America," *The Architectural Record*. Undated.

Whitmeyer, Margaret and Kenn. *Children's Dishes*. Paducah, Kentucky: Collector Books, 1993.

Zim, Herbert S. *Birds, A Guide to the Most Familiar American Birds*. New York: Golden Press, 1956.

Periodicals – Magazines and Catalogs

American Home, 1951, 1952

Blue Ridge Beacon, 1996-1997

Better Homes and Gardens, 1948, 1950, 1951, 1953, 1954

Butler Brothers Catalog, 1952

Chicago Sunday Tribune, August 30, 1942

Country Folk Art, 1992

Country Living, 1993

Erwin Record, "Southern Potteries, Incorporated" Albert L. Price, June 30, 1976.

House Beautiful, 1942, 1947, 1948, 1949, 1950, 1951, 1952, 1953

House and Garden, 1949, 1952

Life, November 7, 1949

Montgomery Ward Catalog, 1942, 1943, 1945, 1951, 1952, 1953, 1954, 1955

National Blue Ridge Newsletter, 1985-1998

Sears and Roebuck Catalog, 1946, 1948, 1949

Spiegel Catalog, 1948

Index

A
ABBY ROSE 49
ABRACADABRA 156
ABUNDANCE 127
ADALYN 62
"ADAM'S FAMILY BOWL" 32
ADORATION 96, 97
ALEENA 69
"ALEENA VARIANT" 99
ALEXANDRIA 94
ALICIA 97
ALLEGRO 98
ALLISON 42, 82
ALOAHA 107
AMANDA 71
AMARYLIS 198
AMERICAN BEAUTY 87
AMHURST 197
ANDANTE 116
ANGELINA 55
ANNADEL 191
ANNALEE 53
"ANNELIESE" 70, 200
ANNE ELAINE 96
ANNIVERSARY SONG 81
ANTIQUE LEAF 153
"ANTIQUE ROSEBUD" 21
"ANTIQUE ROSES" 21
APPALACHIAN GARDEN 102
"APPLE" 135
"APPLESAUCE" 134
APPLE & PEAR 130
APPLE BUTTER 135
APPLE CIDER 36
APPLE CRUNCH 136
APPLE JACK 135
APPLE MIX 130
APPLE STRUDEL 135
APPLE TRIO 133
"AQUA ABBY" 148
"AQUAMARINE DREAM" 17
"ARAB" 168
ARABELLA 68, 123
ARLENE 40, 93
ARLENE WITH LEAF 92
ARLINGTON APPLE 133
"AROUND ROSE MARIE" 49
AROUND ROSEY 147
ARTFUL 106
ART NOUVEAU 24
ASHLAND 69
ASHTRAYS 25
ASTOR BLOSSOM 57
"ATHENA" 29
ATHENS 108
"AUSTIN APPLE" 135
AUTUMN BREEZE 107

B
BABY DOLL 47
BACHELOR BUTTONS 16, 196
BACKYARD BLEEDING HEART 104
BAILEY BLUE 190
BAMBOO 122
"BANANAS & CREAM" 153
"BANANA SWIRL" 35
"BARBIE" 200
BARDSTOWN 106
"BAY APPLES" 134
BEADED APPLE 132, 201
BEADED CHAIN 199
BEADED CHERRY 137
"BECK'S BLOSSOMS" 190
BECKY 64
BEGGARWEED 199
BELLE HAVEN 170
BERRY PATCH 116, 120
BERRYVILLE 139
BETHANY BERRY 135
BETTY 95
BEVERLY 45
"BILL" 117
"B.J." 98
BLACKBERRY LILY 61, 195
BLACKBIRDS 157
BLARNEY 48
BLEEDING HEART 199
BLUE BLOSSOM 97
"BLUE BUTTERFLY" 29
BLUE CURLS 44
"BLUE EYES" 56
BLUE FLOWER 49, 75
BLUE HEAVEN 191
BLUE IRIS 78
BLUE LINE FARM 169
BLUE MOON 97
"BLUE PLUMES" 183
"BLUE POPPY" 36
BLUE TREE 116
BLUE WILLOW 22
BLUEBELL BOUQUET 91, 92, 196, 197
BLUEBONNET 34
"BLUEBONNETS AND POSIES" 37
BLUEFIELD 121
BONAIRE 107
"BOOMERANG" 153
BORDER PRINT 40, 81, 183, 187
"BORDER TEAL LEAF" 153
BOUNTIFUL 128
BOUQUET 42
BOUTONNIERE 105
BRACELET 97
BREATH OF SPRING 58
BRECKENRIDGE 46, 195
BRIAN 107
BRIAR PATCH 52
BRIARWOOD 115
BRIDESMAID 57, 199
BRISTOL BOUQUET 51, 65
BRITTANY 174, 202
BROWN DAISY 109
BROWNIE 92
BUBBLES 152
BUCKETFUL 129
"BUCKY" 147
"BUDDIE'S DREAM" 198
BUG-A-BOO 42, 195
"BURGUNDY BETSY" 89
"BURGUNDY SNOW-FLAKE" 37
BUTTERCUP 198
BUTTERFLY 88
BUTTERFLY & LEAVES 186

C
CADENZA 114
CALAIS 175
"CALHOUN'S BLOSSOMS" 42
CALICO 78, 79, 122
CALIFORNIA POPPY 95
CALLAWAY 102
CAMELOT 114
CANTATA 182
CAPRICIOUS 34
CARETTA CATTAIL 122
CARLILE 55
"CARMEN BETH" 33
CAROL ANN 42
CAROL'S CORSAGE 55
CAROLINA ALLSPICE 112
CAROLINE 110
"CASEY" 122
CASH POTTERY 217, 218, 219, 220, 221
CATOSA 28
CECEILIA 110
CELANDINE 193
CHABLIS 128
"CHALK" 19,20
CHAMBLEE 44, 200
CHAMPAGNE PINKS 58
CHANTICLEER 168
CHARLESTON 109
CHARM HOUSE 85, 86
CHARMER 58
CHATHAM 68
CHEERFUL 72
CHEERIO 112, 113
CHERISH 37
CHEROKEE ROSE 111
CHERRY BLOSSOM 125
CHERRY COBBLER 136
"CHERRY CROSS" 137
CHERRY DROPS 124
CHERRY TREE GLEN 137
CHERUBS 38
CHICKORY 45, 52, 194
CHIFFON 115, 120
CHINTZ 84
CHRISTINE 52
CHRISTMAS DOORWAY 181
CHRISTMAS ORNAMENT 181
CHRISTMAS ROSE 181
CHRISTMAS TREE 181
CHRISTMAS TREE WITH MISTLETOE 181
"CHRISTY" 82
CHRYSANTHEMUM 65, 91
CINNABAR 58
CLAREMONT 102
"CLASSIC GREEN" 38
"CLEARLY STRAWBERRIES" 139
CLOUSE POTTERY 224, 225
CLOVER WREATH 181
CLOVERLAWN 30
"CLUCKERS" 169
COCK-A-DOODLE 168, 169
COCKY LOCKY 168
"COLETTE" 96, 193
"COLLAGE" 33
COLONIAL BIRDS 158
"COLOR RING" 22
COLUMBINE 119
CONASAUGA 79, 85
CONFETTI 100
"CONRO" 112
CORDELE 57
COREOPSIS 187, 201
COSMIC 73
COSMOS 115
COTTON CANDY 36
COUNTRY GARDEN 195
COUNTY FAIR 141
COVINGTON 69
COWSLIP 74
CRAB APPLE 136
CRADLE 166
CREST 18
CRISS CROSS 132
CROCUS 55
CROSS STITCH 184
"CROSS BAR" 147
"CROWNFLOWER" 31
CROWNVETCH 27
CRUSADER SERIES 39
"CRUISING" 155
CUMBERLAND 99
"CURLY" 111
CYCLAMEN 65
CYNTHIANA 199

D
DAISY CHAIN 40, 50, 183
"DAISY CHAIN VARIANT" 40
DAISY GOLD 113
DAMASCUS 109
DANA 128
DANIEL BOONE 164, 165
"DARK CHOCOLATE" 23
"DAY & NIGHT" 148
"DAYLILY" 103
"DEANNA" 48
DEBONAIR 35
"DEEP ROSE" 33
"DELFT DAISY" 34
DELLA ROBBIA 125
DELPHINE 84
DELTA DAISY 192
DEMOREST 46
DESERT FLOWER 115
DISTLEFINK 157
"DIVINITY" 37
DIXIE HARVEST 126
"DODE" 30
DOE EYES 112
DOGTOOTH VIOLET 83, 87
DOLLY 33
"DOMINICKERS" 169
DO-SI-DO 173
"DOTTED CLUCKERS" 167
DOUBLE CHERRY 137
"DOUBLE DAZZLE" 72
DREAM FLOWER 196
DREAMBIRDS 159, 160
DREAMY 36
DUFF SALAD SET 142, 143
DUFFIELD 95
DUNGANNON 118
DUPLICATE 93
DUSTY 150
"DWARF TULIPS" 44

E
"EARTHY WILD CHERRY (#3)" 138
EASTER PARADE 78
"EDDIE" 63
EDGEMONT 57
EDITH 28
EGLANTINE 47
"ELBERTON" 116
ELEGANCE 77, 79
"ELUSIVE" 87
EMALEE 189
EMLYN 101
EMMA 32
"EMPRESS" 89
ENCHANTMENT 63
"ENCORE" 35, 36
ENGLISH GARDEN 46
ERIN 125
ERWIN 63
ERWIN ROSE 80
ERWIN SPRING 84
ERWIN POTTERY 215, 216, 217
EVENTIDE 169
EVERLASTING 49
EXHUBERANT 61
EXOTIC 123

F
FAIRFIELD 70
FALL COLORS 79, 180
"FALL CROCUS" 92
FALLING LEAVES 169, 179
FANCY FREE 49, 197
FARMYARD 170
FASCINATION 74
"FEATHERS" 156
FIELD DAISY 71
FIGURINES 208, 209
FINESSE 30
"FIREFLY" 28
FIRST CLASS 19
FLAMBOYANT 41
FLAMINGO POND 160
FLAPPER 115
"FLATLINE" 20
FLIRT 64, 84
FLOUNCE 58
"FLOW" 37
FLOWER BASKET 18
FLOWER BOWL 55
FLOWER CABIN 163
FLOWER CHILDREN 203
FLOWER RING 61
"FLOWER SONG" 53
"FLOWER TREE" 153
"FLOWERS OF BLUE" 89
FLOWRY BRANCH 48
FOLK ART FRUIT 127
FOLKLORE 171
FONDEVILLE FRUIT 128
FONDVILLE PUNCH 128
FORGET-ME-NOT 195
FOUR CORNER ROSE 88
FOX GRAPE 76
FRAGERIA 139
FREEDOM RING 125
FRENCH KNOTS 121
FRENCH PEASANT 175, 176, 202, 203
FRENCH VIOLETS 151
"FRESH FRUIT" 128
FRIENDSHIP PLAID 151
FRUIT CHILDREN 207
FRUIT COCKTAIL 142
FRUIT CRUNCH 129
FRUIT FANTASY 127
FRUIT MEDLEY 128
FUCHSIA 51
FULL BLOOM 99

G
GAILY 88
GAME COCK 168
GARDEN FLOWERS 59
GARDEN LANE 85, 86
GARDEN PINKS 120
GARDEN WEDDING 37
GARLAND 70
GEORGIA 86, 87
GEORGIA BELLE 53
"GEORGIA PEACHES" 126
GERBER DAISY 96
"GERT" 45
GILBERTINE 83
GILLYFLOWER 26
GINGHAM APPLE 131
GLENDA 34
"GLENN" 106
GLIMMER 156
GLORIA JEAN 106
GLORY 108
'GOLD BAND" 19
"GOLD BRAID" 20
GOLD CABIN 163
"GOLD CHINTZ" 86
"GOLD LEAF" 19
"GOLD LUSTRE" 17
"GOLDEN GIRL" 84
GOLDEN JUBILEE 194
GOOD HOUSEKEEPING ROSE 80
"GOOD HOUSEKEEPING VIOLETS" 82
GORDON VIOLETS 82
"GRACE MARIE" 31
GRANDMOTHER'S GARDEN 76, 184, 194
GRANDMOTHER'S PRIDE 96
GRANNY SMITH APPLE 135
"GRANNY'S PLATTER" 18
GRAPE IVY 118
GRAPE SALAD 202
GRAY DAY 120
GREEN LANTERNS 101
GREEN PEAR 129
GREEN PEPPER 145
"GREEN PLAID" 150
"GREEN WILLOW" 75
GREENBRIAR 105
GREENCASTLE 130
GRESHAM 62
GUMDROP TREE 99
"GUSTY" 108
"GWEN" 98
GYPSY DANCER 29, 57
GYPSY FLOWER 30
GYPSY FRUIT 127

H
HALF & HALF 73
HAM 'N EGGS 167
"HAPPY ELEPHANT" 205
HARDY APPLE 132
HARVESTIME 126
"HARVEY'S FRUIT" 140
HAWAIIAN FRUIT 129
"HAYSTACK" 178
HEIRLOOM 100
"HENRY" 117
HERB WREATH 200
"HERE'S LOOKIN' AT YA" 32
"HERRINGBONE" 35
HEX SIGN 50
HIBISCUS 83
HILLIARY 137
HILLSIDE 171
"HOLLAND" 22
HOLLYBERRY 181
HOLLYHOCK 60
HOMEPLACE 171
"HOMER" 25
HOMESTEAD 171
"HONEY BOUQUET" 28
"HOOVER" 24
"HOPE'S FRUIT" 129
HOPSCOTCH 43
"HUGH KIBLER" 118, 119
HUMPTY 204

I
"INEZ" 72
"IN YOUR DREAMS" 81
INDIAN 166
INGE 116
IRENE 41, 199
IRISH MARY 70, 200

J
"JAMI" 21
JANA 59
JANICE 27
"JAN'S TULIPS" 193
JASON 128
"JEAN'S DELIGHT" 52, 200
"JENNIFER" 116
JENNY WREN 166
JESSAMINE 86, 102, 103
JESSE 49
JESSICA 43
JIGSAW 203
JOE'S APPLES 136
JOHN'S PLAID 150
"JONATHAN'S MEMORY" 75
JOSEY'S POSIES 60
JUBILEE FRUIT 140, 141
JUDITH 69
"JUGGLING" 147
JUNE APPLE 133
JUNE BOUQUET 65, 71, 83
JUNE BRIDE 54
"JUNE TULIPS" 24

K
KANSAS GAY FEATHER 60
KAREN 65, 66, 197, 201
KATE 64
"KATE'S ROSE GARDEN" 46
"KATHLEEN" 113
"KATIE" 25
"KAYE" 56
KAYLA RUTH 93
KELCI 53
KELLY 93
KELVIN 36
KENNESAW 67
KIBLER 105
"KIND OCTOBER" 109
KING'S RANSOM 65
KISMET 117
"KRIS LEIGH" 48

L
LACE LEAF COREOPSIS 72
"LACY RADIANCE" 31
"LADIES IN WAITING" 38
LADY MOUSE 206
"LANA" 106
LANGUAGE OF FLOWERS 100
LARGO 83
LAURA 55, 65
LAUREL BLOOMERY 26
LAUREL WREATH 67
"LARRY'S LUSTRE" 32
LAVENDER FRUIT 126, 201
LAVENDER IRIS 59
"LAZY DAY FARMS" 172
LE COQ 175
LE COQ SOLEIL 175
LEA MARIE 67
LEAF & BAR 185
LEAF & CIRCLE 152
LEAF 153, 183, 186
LEAF SPRAY 113
"LEAFY FRUIT RING" 130

LEAVES OF FALL 119
"LEMON YELLOW" 124
LIGHTHEARTED 91
"LIME ICE" 152
"LIMEY" 152
"LINDA" 152
LITTLE GIRL 90
LIZZIE'S GIFT 92
"LOGO" 27
LORETTA 150
LOTUS 104
LOUISA 63
LOVE SONG 173
"LUCINDA" 18
LUELLA 100
"LUSTRE GRAPES" 22
"LUSTRE WILLOW" 22
"LYDIA" 154
LYONNAISE 174, 175

M

"MADISYN'S ROSES" 32
MADRAS 154
MADRIGAL 52
"MAGGIE LEE" 79
MAGIC CARPET 196
MAGNOLIA 103
MALLARD 89, 178
MAPLE LEAF RAG 100
MARDI GRAS 184, 191
MARIE'S POTTERY 223
"MARIE'S ROSE" 88
MARIETTA 110
"MARIGOLD" 18
MARINER 155, 186
"MARINER VARIANT" 155
"MARION" 43
MARJORIE 91
MARK 109
MAROON PLAID 151
MARTHA PITCHERS 143, 146, 148, 152
"MARTHA'S EGGS" 84
MARY 124, 187
"MARY ANN" 106
"MATILDA" 30
MAYFLOWER 121, 150
MAYTIME 44
MAZURKA 103
MEADOW BEAUTY 68
MEADOWLEA 114
MEAGAN 66
MELODY 79, 80, 83
"METALLICA BLUE" 27
MEXICANO 172
MEYLINDA 65
"MICHAEL'S ROSE" 88
MICHELLE 95
MIDAS TOUCH 113
"MIDNIGHT" 78, 87
MIDSUMMER ROSE 29
MIMIC 122
MING TREE 123
"MINNIE" 80
MIRROR IMAGE 64
MIRROR, MIRROR 72
MISHA 75
"MISS LIME" 89
MISS MOUSE 206
MISS PIGGY 204
MISSISSIPPI 117
MISTRESS MARY 198
MOCASSIN 99
MOD OAK LEAF 179
MOD TULIP 199
"MOD VARIANT" 50
MODERN LEAF 117
"MONOGRAM" 19
"MONTY" 40
MOOD INDIGO 84
"MOSS ROSE VARIANT" 106
"MOTLEY BLUE" 80
MOTTO WITH SNOW-FLAKE 27
MOUNTAIN ASTER 130
MOUNTAIN BELLS 53
MOUNTAIN CHERRY 137
MOUNTAIN CRAB 201
MOUNTAIN DAISY 72
MOUNTAIN GLORY 73, 74, 122
MOUNTAIN IVY 99
"MOUNTAIN LAUREL" 73
MOUNTAIN MEADOWS 97
MOUNTAIN NOSEGAY 81
MULTICOLOR TULIP TRIO 68
MURIEL 189
"MUSEUM PIECE" 61
"MYSTIC MOSAIC" 149

N

NAÏVE 30
"NANA" 20
"NASCO PRINCESS" 89
"NASCO ROSES" 183, 184
NASSAU 45
NAUGHTY 88
NEEDLEPOINT FRUIT 130
NESTING BIRDS 156
"NEW WORLD" 22
NIGHT FLOWER 113
NIGHTLIGHT 31
NOCTURNE 198
NOCTURNE (RED) 64
"NOCTURNE (RED) VARIANT" 199
NONSENSE 43
NORMA 62
NORMANDY 175
"NORTHERN DOGWOOD" 110
"NOVELLA" 20
NOVE ROSE 76, 85, 194

O

OAKDALE 101
OBION 94
OCTOBER BLUE 34
"ODD COUPLE" 132
"OLD ROSE" 17
OLGA 63
ORANGE QUEEN 110
ORCHARD GLORY 129
ORCHID 109
"O'ROSIE" 49
"OUCH" 75
"OUR LISA" 65

P

"PAINTER'S PAIN" 23
"PAM" 31
PANDORA 62
PANSY TRIO 67
"PARABOLA" 155
PÅRROT JUNGLE 161
PARTRIDGE BERRY 114
"PASADENA" 90
PASQUE TULIP 86
PASTEL LEAF 154
"PASTEL TULIP CIRCLE" 99
"PATTY" 64
PAUL REVERE 165
PAULA'S LEI 90
"PEACH POPPY" 20
PEACOCK 87, 159
"PEARL" 23
PECAN 148
PEGGIE'S POSIES 98
"PEGGIE'S POSIES VARI-ANT" 99
"PEGGY SUE" 26
PENNY SERENADE 58
PEONY 46, 66, 196
PERFORMING SEAL 205
PETAL POINT 179, 180, 181
PETER'S HOT CHOCOLATE 24
"PETIE" 191
PETIT POINT 192
PETITE FLOWER SET 68
PETITE FRUITS 141, 142
PETUNIA 67, 193
PHLOX 44
PIANISSIMO 112
PIEDMONT PLAID 152
PIGGY BLUES 204
PILGRIMS 178
PINCUSHION 114
"PINK CHAMPAGNE" 120
"PINK CROCUS" 91
PINK DAISY 88
"PINK LUSTRE" 23
"PINK MORNING GLORY" 60
PINK PUPPY 204
PINKIE LEE 108
PIONEER WOMAN 164
PIPPA 126
PIXIE 85
PLANTATION IVY 118
PLUME 197
POINSETTA 179, 180, 191
POLKA DOT 57
"POINTER" 182
POM POM 97
"POM POM VARIANT" 192
POND IRIS 21, 186
POPPY DUET 95
"POP'S RIBBON ROSE" 46
"POWDER PUFFS" 88
PRACTICAL PIG 206
PRARIE ROSE 40, 200
PRECIOUS 93
"PRECIOUS PEAR" 129, 130
PRETTY PETALS 95
PRIM 111
PRIM POSIES 53
"PRISSY" 67
PROVINCIAL FARM SCENES 174
"PUNY POSIES" 32
PURITAN 60
PURPLE CROWN 60
PURPLE MAJESTY 59
"PYRACANTHA VARIANT" 117

Q

QUAIL 161, 162
QUAKER APPLE 133
QUARTET 153
"QUEEN" 21
"QUILTED APPLE VARIANT" 131
QUILTED FRUIT 131
QUILTED IVY 122
"QUILTED STRAWBERRY" 131

R

RADIANCE 53
"RADISHES" 145
RAINY DAY 151
RAZZLE DAZZLE 108
RECOLLECTION 44, 84
RED APPLE 134, 201
RED BANK 90
RED CARPET 94
"RED DAISY" 66
"RED GAIETY" 64
RED HILL 70
RED LEAF 203
"RED PETUNIA" 193
RED RING 103
RED ROOSTER 169
RED STAR 71
RED TULIP 97, 98, 184, 188
REFLECTION 26, 76
REHOBOTH 43
REMEMBRANCE 52, 201
RHAPSODY 62
RIBBON PLAID 151
RICHARD 193
RIDGE DAISY 73, 84
RIDGE ROSE 66, 67
"RINGED LEAF" 154
"RINGED FRUIT" 126
"RINGED IVY" 117
ROAN MOUNTAIN ROSE 53
ROBERT E. LEE 23
ROBIN 119
"ROBY" 205, 206
ROCK ROSE 56, 185
"ROCK ROSE VARIANT" 56
ROCKPORT ROOSTER 169
ROCKY PLAID 150
ROMANCE 77, 78, 194
"ROMEO & JULIET" 39
ROOSTER 170
ROOSTER MOTTO 168
"ROPE 'EM, COWBOY" 172
"ROSALIE" 185
"ROSARIO" 18
ROSE GARDEN 86
ROSE MARIE 87, 194
ROSE OF SHARON 47, 78, 80, 81, 84, 85
ROSE RED 93
ROSEANNA 47
ROSE MARIE 77, 79, 87
ROSEMARY 94, 194
"ROSE WREATH" 20
ROSY FUTURE 195
ROUNDELAY 26, 35, 148
"ROYAL WREATH" 19
RUE DE LA PAIX 121
"RUFFLES" 206
RUGOSA 73
"RUTH" 36
RUTH ANNA 58, 64

S

"SABRE" 74
SAILBOAT 155
SAILFISH 160
"SALEM" 75
SALMAGUNDI 16
SANDRA 152
"SANKIE" 71
SANTA MARIA 23
SARASOTA 150
SAREPTA 63
SCARLET LEAVES 107
"SCONCES" 19
SECRET GARDEN 105
SHANNON 46
"SHEILA" 50
SHELLING PEAS 167
SHENANDOAH 41
"SHEENA" 23
SHEPHERD'S PURSE 118
SHEREE 58
SHERRY 70
SHIMMER 120
SHOO FLY 74
SHOW OFF 26
"SIDE BY SIDE" 44
"SILLY CLOWN" 205
"SIMPLE FRUIT" 130
"SIMPLY WILD CHERRY (#1)" 138
SINGLETON 137
"SISTERS" 80
SKEETER 189
SKITTER 47, 87, 190
SKYLINE SONGBIRDS 159
SMOKEY MOUNTAIN LAUREL 146
SNIPPET 54, 188
"SO BLUE" 148
"SO PINK" 149
SOLIDAGO 118
"SONDRA" 60
SONG SUNG BLUE 198
SONGBIRDS 157
SONORA 171
SOUTHERN CAMELLIA 104
SOUTHERN DOGWOOD 115
SOUTHERN ROSE 45
SOUTHERN RUSTIC 148
SOUTHERN STARBURST 107
"SPACE" 149
"SPIDER VIOLETS" 110
SPIDERWEB 26, 37, 110, 148, 149, 150
SPINDRIFT 36, 183
SPIRAL PITCHERS 147
SPLASH 33
"SPLASHY SOUTHERN ROSE" 30
"SPLATTER ME BLUE" 26, 27
"SPOT" 24
"SPRAY AROUND" 113
"SPRIG VARIANT" 119
SPRING BEAUTY 55
SPRING BLOSSOM 102, 103
SPRING BOUQUET 196
SPRING GLORY 25
SPRING MORNING 101
SPRING SONG 41
SQUARE DANCERS 172, 173
"SQUIGGLE" 76, 100
STACCATO 26, 71, 121, 205
"STAMPED PANSIES" 34
STANHOME IVY 119
STARDANCER 56
STARFLOWER 120, 201
STENCIL 151
STEPHANIE 82
STETSON POTTERY 111, 114
STRAW HAT 167
STRAWBERRY DUET 140
STRAWBERRY GARDEN 140
STRAWBERRY PLANT 140
STRAWBERRY SUNDAE 139
"STREAKERS" 169
"ST. PATRICK'S FLOWER" 99
SUE-LYNN 49
SUMMER SUN 50
SUMMERSWEET 196
SUMMERTIME 76
SUN BOUQUET 71, 105, 114
"SUN HAT" 167
SUNBURST 26, 114
SUNDAY BEST 65
SUNGOLD 25, 96
SUNNY SPRAY 112
SUSAN 123
SUTHERLAND 128
SWAMP MALLOW 106
"SWEET BOUQUET" 183
SWEET PEA 87, 199
SWEET ROCKET 122
SWEET STRAWBERRIES 139
SWEETHEART 139
SWEETIE PIE 135
SWISS DANCERS 174
SYDNEY'S HOPE 48
SYMPHONY 196

T

TAFFETA 69
TALISMAN LINE 160
TANSY 51
"TARA'S ROSE" 56
"TARGET" 31
TARTAN APPLE 131
TARTAN FRUIT 132
TEA ROSE 109
TEAL ROSEANNA 190
"TEAL RUSTIC" 148
"TEENSY" 45
TEMPO 104
"TENNESSEE PINE" 114
TENNESSEE WALTZ 52
TERESA 59
"TERESIA'S JUG" 151
"TERRI ANN" 48
TEXAS ROSE 72
THANKSGIVING TURKEY 178, 179
"THE PRIZE" 47
"THE TREE" 31
THREE LIITTLE PIGS 207
"TICKLE" 197
TIGER EYE 43
TIGER LILY 61
TINY 108
TOGETHER 90
"TOMORROW" 63
TOP KNOT 69
"TRACKS" 147
TRIBUTE 50
"TRIED & TRUE" 17
TRIFLE 29
TRIPLET 40
TROPICAL 170
TRUMPET VINE 102
TULIP 54, 88, 183
TULIP BUDS 190
TULIP CORSAGE 48
TULIP GARDEN 98
TULIP RING 25
TULIP ROW 68
"TULIP SCROLLS" 98
TULIP TREAT 98
TUNA SALAD 155
TURKEY GOBBLER 162
TURKEY WITH ACORNS 178, 179
"TURQUOISE CHAIN" 38
TWEET 156
TWIG 106, 107
TWIN TULIPS 99, 109
"TWIRLY APPLE" 133
"TWISTED RIBBONS" 151
TWO OF A KIND 26, 68
"TWO OF A KIND VARIANT" 69
"TYLER" 114, 121

U

"UPSIDE DOWN APPLE" 132
UPSTART 71
"URN OF ROSES" 27

V

VALDOSTA 111
VALENTINE ASTERS 51
"VALLEY BLOSSOM" 48, 56
VANITY FAIR 74
VARIETY 68
VEGETABLE PATCH 145
VEGETABLE SOUP 145
"VEGETABLE STEW" 145
VEGGI 144
VELMA 102
"VENUS" 108
VERONA 75
VIBRANT 98
VICKY 32
VICTORIA 183
VINCA 187
"VINCA VARIATION" 26
VIOLA 56
VIOLET 55
"VIOLET BLUSH" 112
VIOLET CIRCLE 55
VIOLETS FOR JEAN 54
VONDA 192

W

WALTZ TIME 77
"WANDA" 79
WATERLILY 88, 91, 190
"WATERMELON SLICE" 153
WEATHERVANE 170
WEATHERVANE COCK 168
WEE LEAF 147
WHIG ROSE 89
WHIMSEY 75
WHIP TAC 134
WHIPSTITCH 133
WHIRL 117
WHIRLIGIG 62
"WHIRLY-WHIRLY" 61
"WHISPER" 80
WILD GERANIUM 111
WILD IRISH ROSE (RED) 65
WILD IRISH ROSE (YEL LOW) 187
WILD MORNING GLORY 189
"WILD PETUNIA" 60
WILD ROSE 64
WILD STRAWBERRY 64, 137, 138
WILD TURKEY 163
WILDWOOD 48
WILDWOOD FLOWER 43, 92, 200
"WILLENE" 92
WILLOW 118
WINDFLOWER 64
"WINNIE'S MOUNTAIN DAISY" 112
"WINTER PANSIES" 60
WINTERTIME 182
WITCHERY 76
WRINKLED ROSE 54, 189

Y

"YALLISON" 42
YELLOW BUNNY 203
YELLOW DREAMS 96
YELLOW MUMS 59
YELLOW NOCTURNE 74, 184, 192
YELLOW PANSY 59
YELLOW PLUME 51, 188
YELLOW RABBIT 203
"YELLOW RAMBLER" 51
YELLOW RIBBON 47
YELLOW ROSE 82
YELLOW TEA ROSE 33
"YOLANDA" 33
YORKTOWN 160